APPLIED
STOCHASTIC
PROCESSES

APPLIED
STOCHASTIC
PROCESSES

MING LIAO

AUBURN UNIVERSITY
ALABAMA, USA

CRC Press
Taylor & Francis Group
Boca Raton London New York

CRC Press is an imprint of the
Taylor & Francis Group, an **informa** business
A CHAPMAN & HALL BOOK

CRC Press
Taylor & Francis Group
6000 Broken Sound Parkway NW, Suite 300
Boca Raton, FL 33487-2742

First issued in paperback 2019

© 2014 by Taylor & Francis Group, LLC
CRC Press is an imprint of Taylor & Francis Group, an Informa business

No claim to original U.S. Government works

ISBN-13: 978-1-4665-8933-9 (hbk)
ISBN-13: 978-0-367-37977-3 (pbk)

Contents

Preface

The purpose of these notes is to present a concise account of applied stochastic processes as usually covered in a first-year graduate course, emphasizing applications and practical computation, but also developing an essentially complete mathematical theory. The topics, after a quick review of basic probability, include Poisson processes, renewal processes, discrete time and continuous time Markov chains, Brownian motion, and an introduction to stochastic differential equations. The main applications are queues, but other examples are also considered, including the mathematical model of a single stock market. We have tried to present the materials in a quick to-the-point fashion, possibly with some motivation and short examples, but without much digression. The mathematical theory is developed with strong emphasis on probability intuition, often informally, to be easily accessible, but always based on sound reasoning. Most sections end with a few closely related exercises. The solutions to all exercises are included in a separate solutions manual.

The short bibliography contains the books and papers used in the preparation of the notes. It is not intended to be comprehensive, and so does not include many other good books on applied stochastic processes.

Students of my stochastic process classes endured the early versions of the notes, and provided useful inputs. I wish to thank the anonymous reviewers for the helpful comments and my colleague Erkan Nane for providing a list of errors. A lot of effort has been put into these notes, and my wife's support and understanding are always important for me to complete such a project.

MATLAB and Simulink are registered trademarks of The Math-Works, Inc. For product information, please contact:

The MathWorks, Inc.
3 Apple Hill Drive
Natick, MA 01760-2098 USA
Tel: 508 647 7000

Fax: 508-647-7001
E-mail: info@mathworks.com
Web: www.mathworks.com

Chapter 1

Probability and stochastic processes

The first eight sections summarize basic concepts and facts in probability theory with little explanation and very few proofs. The reader may consult a standard graduate text on probability, such as [2], for more details. The stochastic processes, stopping times, and conditional expectation are introduced in the last three sections.

1.1 Probability

Sample space and events: The set of all possible outcomes is called the sample space and is denoted by Ω. A collection of subsets of Ω, including Ω itself and the empty set $\emptyset$, are called events. This will be defined more precisely in the note on the measure theory later, but a reader who is not familiar with the measure theory may just assume any subset of Ω is an event. We will assume the reader is familiar with the basic set notation and operations such as the membership $\in$, the inclusion $\subset$, the union $\cup$, the intersection $\cap$, and the complement E^c of an event E.

Probability: The probability $P(E)$ is a function defined for every event E, which satisfies the following three basic axioms (or rules):

(a) $0 \le P(E) \le 1$.
(b) $P(\Omega) = 1$ and $P(\emptyset) = 0$.
(c) $P(\cup_{n=1}^{\infty} E_n) = \sum_{n=1}^{\infty} P(E_n)$ for any sequence of disjoint events $E_1, E_2, E_3, \ldots$,

Property (c) is called the countable additivity of the probability as a sequence of objects is called a countable set in mathematics.

The sample space Ω together with a probability P is called a probability space. The following properties are easy consequences of the above three axioms:

(d) If $A \subset B$, then $P(A) \leq P(B)$.
(e) $P(A^c) = 1 - P(A)$.
(f) $P(A \cup B) = P(A) + P(B) - P(A \cap B)$ for two events A and B.

Note on measure theory: For readers who know measure theory, it should be mentioned that the probability P on Ω is a measure, and it may not be defined on all subsets of Ω, but only on a collection of subsets which form a σ-algebra $\mathcal{F}$ on Ω. Thus, only subsets contained in $\mathcal{F}$ are called events.

By definition, a σ-algebra on Ω is a collection $\mathcal{F}$ of subsets of Ω which contains Ω and $\emptyset$, and is closed under taking complement and countable union in the sense that if $A \in \mathcal{F}$, then $A^c \in \mathcal{F}$, and if $A_n \in \mathcal{F}$ for $n = 1, 2, 3, \ldots$, then $\cup_{n=1}^{\infty} A_n \in \mathcal{F}$. It then follows that a σ-algebra is also closed under taking countable intersection. On the real line $\mathbb{R}$, the smallest σ-algebra containing all intervals is called the Borel σ-algebra on $\mathbb{R}$ and is denoted as $\mathcal{B}(\mathbb{R})$. We may also consider the Borel σ-algebra $\mathcal{B}(I)$ on an interval I, defined as the smallest σ-algebra on I containing all subintervals of I.

The space Ω together with a σ-algebra $\mathcal{F}$ is called a measurable space. A function m defined on $\mathcal{F}$ is called a measure on Ω if $m(A) \geq 0$ for $A \in \mathcal{F}$, $m(\emptyset) = 0$ and $m(\cup_n A_n) = \sum_n m(A_n)$ for any sequence of disjoint $A_n \in \mathcal{F}$. The triple $(\Omega, \mathcal{F}, m)$ is called a measure space. A probability P on Ω is a measure satisfying $P(\Omega) = 1$, called a probability measure. There is a unique measure m on $\mathbb{R}$ equipped with the Borel σ-algebra $\mathcal{B}(\mathbb{R})$ such that $m((a, b]) = b - a$ for all $a < b$, called the Lebesgue measure on $\mathbb{R}$.

Increasing and decreasing sequences of events: A sequence of events A_n, $n \geq 1$, are called increasing if $A_n \subset A_{n+1}$ for all $n \geq 1$. In this case, $A = \cup_n A_n$ is called the limit of A_n, denoted as $A_n \uparrow A$. Similarly, A_n are called decreasing if $A_n \supset A_{n+1}$. Then $A = \cap_n A_n$ is called the limit of A_n, denoted as $A_n \downarrow A$.

It can be shown that if $A_n \uparrow A$ or $A_n \downarrow A$, then

$$P(A) = \lim_{n \to \infty} P(A_n). \tag{1.1}$$

Conditional probability: Let A and B be two events with $P(B) > 0$.

The conditional probability of A given B is defined by

$$P(A \mid B) = \frac{P(A \cap B)}{P(B)}. \tag{1.2}$$

Note that for a fixed B, $P_B(A) = P(A \mid B)$ is a probability as A varies over all events, which may be different from $P(A)$. Note also $P_B(A \mid C) = P(A \mid B \cap C)$, which may be denoted as $P(A \mid B, C)$.

Total probability law: Let B_n be a sequence of events that form a partition of the sample space Ω, that is, B_n are disjoint and $\cup_n B_n = \Omega$. Then for any event A,

$$P(A) = \sum_n P(B_n)P(A \mid B_n). \tag{1.3}$$

Independence: If the conditional probability of an event A given another event B is the same as the (un-conditional) probability of A, that is, if $P(A \mid B) = P(A)$, then by (1.2) it is easy to see that

$$P(A \cap B) = P(A)P(B). \tag{1.4}$$

Two events A and B are called independent if (1.4) holds.

More generally, a collection of events are called independent if for any finitely many events $A_1, A_2, \ldots, A_n$ in the collection,

$$P(A_1 \cap A_2 \cap \cdots \cap A_n) = P(A_1)P(A_2) \cdots P(A_n). \tag{1.5}$$

Note that if the events in a collection are just pairwise independent, that is, if any two events A and B in the collection satisfy (1.4), then the collection may not be independent.

1.2   Random variables and their distributions

Random variables: A random variable X is a real-valued function on the sample space Ω. In a more formal discussion, the sample space Ω is always equipped with a σ-algebra $\mathcal{F}$ of events and a random variable X is assumed to be measurable under $\mathcal{F}$, or $\mathcal{F}$-measurable, in the sense that $[X \leq r] \in \mathcal{F}$ for all real r. This is equivalent to $[X \in J] \in \mathcal{F}$ for any interval J, and is also equivalent to $[X \in B] \in \mathcal{F}$ for any Borel subset B

of $\mathbb{R}$, that is, any $B \in \mathcal{B}(\mathbb{R})$. We will not emphasize the measurability of a random variable, however, it will be clear that the probability computation of a random variable requires the measurability.

Distribution function: The probability distribution function, or simply the distribution function, of a random variable X is defined by

$$F(x) = P(X \le x). \tag{1.6}$$

It may also be denoted as $F_X(x)$ when the random variable X needs to be indicated. It has the following properties:

(a) $0 \le F(x) \le 1$;
(b) $F(x)$ is increasing in the sense that $F(x_1) \le F(x_2)$ for $x_1 < x_2$, and is right continuous in the sense that $F(x) = F(x+)$, where $F(x+) = \lim_{x < y \to x} F(y)$ (right-hand limit at x);
(c) $\lim_{x \to \infty} F(x) = 1$ and $\lim_{x \to -\infty} F(x) = 0$.

Almost sure equality: Two random variables X and Y are said to be equal almost surely if $P(X = Y) = 1$. This may be written as $X = Y$ almost surely or simply $X = Y$ a.s. In this case, X and Y have the same distribution, that is, $F_X(x) = F_Y(x)$, which is denoted as $X \stackrel{d}{=} Y$. Note that $X \stackrel{d}{=} Y$ does not imply $X = Y$ a.s. in general.

Existence of a random variable: Any function $F(x)$ satisfying (a), (b) and (c) above is called a distribution function. Given an arbitrary distribution function $F(x)$, there is a random variable X having $F(x)$ as its distribution function, that is, (1.6) holds. Indeed, one may take $\Omega = [0, 1]$ equipped with the Lebesgue measure and for $\omega \in \Omega$, let

$$X(\omega) = \inf\{x \in \mathbb{R};\ F(x) > \omega\},$$

where $\inf\{\cdots\}$ is the infimum or the greatest lower bound of the numbers in $\{\cdots\}$ (defined to be ∞ if the set $\{\cdots\}$ is empty). Note that when $F(x)$ is continuous and strictly increasing for $-\infty < x < \infty$, $X(\omega)$ is just the inverse function $F^{-1}(\omega)$ of $F(x)$.

Discrete random variables and probability mass functions: A random variable X is called discrete if it has only finitely many or countably many possible values. When X has finitely many values, its distribution function $F(x)$ is a step function which steps up at each possible value with the size of the jump being the probability of this value. For a general discrete random variable, $F(x)$ is the sum of the probabilities of all possible values $\leq x$.

The pmf (probability mass function) of X is defined by

$$p(x) = P(X = x). \tag{1.7}$$

Note that $p(x) > 0$ only when x is a possible value of X, and

$$F(x) = \sum_{y \leq x} p(y). \tag{1.8}$$

Continuous random variables and probability density functions: A random variable X is called continuous if its distribution function $F(x)$ is continuous and has a derivative $f(x) = F'(x)$ such that

$$F(x) = \int_{-\infty}^{x} f(y)dy. \tag{1.9}$$

The function $f(x)$ is called the pdf (probability density function) of X. Note that for a continuous random variable X, $P(X = x) = 0$ for any real x.

Deficient random variables: Occasionally we may allow a random variable X to take ∞ or $-\infty$ as a possible value. Such a random variable is called deficient because $P(X \in \mathbb{R}) < 1$. The distribution function, pmf and pdf for a deficient random variable are still defined as before, but their properties need to be suitably modified. For example, $\lim_{x \to \infty} F(x) \leq 1$ instead of $= 1$. In the sequel, a random variable is assumed to be non-deficient unless when explicitly stated otherwise.

More general types of random variables: A random variable does not have to be either discrete or continuous. In general, its distribution function may increase continuously over some interval and also have jumps at other places.

The term random variable may also be used to call any measurable function X on Ω taking values in a general metric space S, which is equipped with a distance and so the notion of convergence is defined. The measurability of X is defined under the Borel σ-algebra $\mathcal{B}(S)$ on S, which is naturally induced by the distance, but whose precise definition will not be given here. For example, several real-valued random variable $X_1, X_2, \ldots, X_n$ together may be regarded as a single random variable $X = (X_1, X_2, \ldots, X_n)$ taking values in the n-dimensional Euclidean space $\mathbb{R}^n$.

In the sequel, a random variable is understood to be real valued, unless when explicitly stated otherwise.

1.3   Mathematical expectation

Riemann-Stieljes integrals: For any two real valued functions f and g on $\mathbb{R}$ with g being right continuous, the Riemann-Stieljes integral over a finite half-open interval $(a, b]$ is defined as follows: Let $x_0 = a < x_1 < x_2 < \cdots < x_n = b$ be a partition of the interval with mesh $\delta = \max_{1\leq i\leq n}(x_i - x_{i-1})$, and let $\xi_i \in (x_{i-1}, x_i]$ for $1 \leq i \leq n$. Then

$$\int_{(a,b]} f(x)dg(x) \ \text{ or } \ \int_a^b f(x)dg(x) = \lim_{\delta\to 0}\sum_{i=1}^n f(\xi_i)[g(x_i) - g(x_{i-1})],$$

provided the limit exists along any sequence of partitions with $\delta \to 0$, and is independent of choices of ξ_i's.

The Riemann-Stieljes integral over the closed interval $[a, b]$ is defined as

$$\int_{[a,b]} f(x)dg(x) = f(a)[g(a) - g(a-)] + \int_a^b f(x)dg(x),$$

where $g(a-)$ is the left-hand limit of $g(x)$ at $x = a$. Note that the jump of $g(x)$ at $x = a$ affects $\int_{[a,b]} f(x)dg(x)$, but not $\int_a^b f(x)dg(x)$.

When $g(x)$ is continuous at $x = a$, the two integrals $\int_{[a,b]} f(x)dg(x)$ and $\int_a^b f(x)dg(x)$ are the same. Note that if $g(x)$ has a continuous derivative $g'(x)$, then

$$\int_a^b f(x)dg(x) = \int_a^b f(x)g'(x)dx.$$

In particular, when $g(x) = x$, the Riemann-Stieljes integral becomes the usual Riemann integral $\int_a^b f(x)dx$ as defined in calculus.

The integral over an infinite interval is defined as a limit, for example, $\int_a^\infty f(x)dg(x) = \lim_{b\to\infty}\int_a^b f(x)dg(x)$, provided the limit exists. Note that

$$\int_{-\infty}^\infty f(x)dg(x) = \int_{-\infty}^0 f(x)dg(x) + \int_0^\infty f(x)dg(x)$$

provided both $\int_{-\infty}^{0} f(x)dg(x)$ and $\int_{0}^{\infty} f(x)dg(x)$ exist and are finite. This even holds when either $\int_{-\infty}^{0} f(x)dg(x)$ or $\int_{0}^{\infty} f(x)dg(x)$, or both, are infinite as long as the sum makes sense.

Expectation: The expectation or expected value of a random variable X is defined by

$$E(X) = \int_{-\infty}^{\infty} xdF(x) = \begin{cases} \sum_x xp(x) & \text{if } X \text{ is discrete} \\ \int_{-\infty}^{\infty} xf(x)dx & \text{if } X \text{ is continuous,} \end{cases}$$

(1.10)

which is also called the mean of X and denoted by μ or μ_X. Note that $E(X)$ may be infinite ($\pm\infty$) or it may not exist. If $X \geq 0$, then $E(X)$ always exists but may be ∞.

Properties: (a) $E(c) = c$ for any constant c;
(b) $E(aX + b) = aE(X) + b$ assuming $E(X)$ is finite, where a and b are constants;
(c) $E(X \pm Y) = E(X) \pm E(Y)$ assuming $E(X)$ and $E(Y)$ are finite.
    Property (a) is obvious; a little effort is needed to prove (b) and (c), but they follow immediately from the formal definition of expectation discussed below.

Lebesgue integrals and expectation: For a measurable function f on a measure space $(\Omega, \mathcal{F}, m)$, the Lebesgue integral $\int_{\Omega} f dm = \int_{\Omega} f(x)m(dx)$ may be defined. For the Lebesgue measure on an interval $I = (a, b)$ equipped with the Borel σ-algebra $\mathcal{B}(I)$, the Lebesgue integral is the usual Riemann integral $\int_{a}^{b} f(x)dx$ for any bounded function $f(x)$ with countably many discontinuous points, and it may also be defined for a function that is measurable under the Borel σ-algebra, called a Borel function. We will not delve into the general theory of Lebesgue integrals. A reader who is not familiar with this subject just needs to know that a Lebesgue integral behaves just like a usual Riemann integral. For example, $\int_{\Omega} af(x)m(dx) = a \int_{\Omega} f(x)m(dx)$ for a constant a, and $\int_{\Omega}(f + g)dm = \int_{\Omega} f dm + \int_{\Omega} g dm$ for two functions f and g on Ω.

The expectation of a random variable X may be defined more formally as

$$E(X) = \int_{\Omega} XdP.$$

(1.11)

This agrees with the definition given earlier.

Expectation of a function of a random variable X:

$$E[g(X)] = \int_{-\infty}^{\infty} g(x)dF_X(x) \tag{1.12}$$

$$= \begin{cases} \sum_x g(x)p_X(x) & \text{if } X \text{ is discrete} \\ \int_{-\infty}^{\infty} g(x)f_X(x)dx & \text{if } X \text{ is continuous,} \end{cases}$$

for any Borel function $g(x)$, assuming the integral in (1.12) exists.

Variance of a random variable X: $\mathrm{Var}(X) = E[(X-\mu)^2] = E(X^2) - \mu^2$, also denoted by σ^2 or σ_X^2, assuming $E(X^2) < \infty$. This provides a measurement of the dispersion of X from its mean μ.

Expectation of a nonnegative random variable X:

$$E(X) = \int_0^{\infty} \bar{F}_X(x)dx, \tag{1.13}$$

where $\bar{F} = 1 - F$. If X takes only nonnegative integer values, then this becomes

$$E(X) = \sum_{n=0}^{\infty} P(X > n) = \sum_{n=1}^{\infty} P(X \geq n). \tag{1.14}$$

Note that (1.13) and (1.14) hold even when $E(X) = \infty$.

The identity (1.13) may be proved by an integration by part, or more formally as follows.

$$E(X) = \int_{\Omega} X dP = \int_{\Omega} \int_0^X dx dP = \int_0^{\infty} \int_{[X>x]} dP dx$$

$$= \int_0^{\infty} P(X > x)dx = \int_0^{\infty} \bar{F}(x)dx.$$

Note that $P(X \geq x) \neq P(X > x)$ for at most countably many x.

1.4   Joint distribution and independence

Joint distribution: The joint distribution function of two random variables X and Y is defined by

$$F(x,y) = P(X \leq x, Y \leq y). \tag{1.15}$$

In relation to the joint distribution, the individual distributions of X and Y are called marginal distributions, which may be obtained from the joint distribution as

$$F_X(x) = F(x, \infty) = \lim_{y \to \infty} F(x, y) \quad \text{and} \quad F_Y(y) = F(\infty, y).$$

Joint pmf and pdf: The joint pmf (joint probability mass function) of two discrete random variables X and Y is the function $p(x, y) = P(X = x, Y = y)$.

Two random variables X and Y are said to have a continuous joint distribution, or to be jointly continuous, if there is an integrable function $f(x, y) \geq 0$ on the plane $\mathbb{R}^2$, called the joint pdf (joint probability density function), such that

$$F(x, y) = \int_{-\infty}^{x} \int_{-\infty}^{y} f(u, v) dv du. \tag{1.16}$$

Then both X and Y are continuous with marginal pdf's given by

$$f_X(x) = \int_{-\infty}^{\infty} f(x, y) dy \quad \text{and} \quad f_Y(y) = \int_{-\infty}^{\infty} f(x, y) dx.$$

Moreover, for any (Borel) subset B of $\mathbb{R}^2$,

$$P[(X, Y) \in B] = \int \int_{B} f(x, y) dx dy.$$

Note that in general, two continuous random variables may not be jointly continuous. For example, if X is continuous and $Y = X$, then X and Y are not jointly continuous because $P(X = Y) = 1$.

Independence: Two random variables X and Y are called independent if for any real numbers a and b, the two events $[X \leq a]$ and $[Y \leq b]$ are independent, that is,

$$P(X \leq a, Y \leq b) = P(X \leq a)P(Y \leq b).$$

This is equivalent to

$$F(x, y) = F_X(x)F_Y(y) \quad \text{for all real } x \text{ and } y.$$

When X and Y are jointly continuous, this is also equivalent to $f(x, y) = f_X(x)f_Y(y)$.

Independence of more than two random variables: The joint distribution and independence may be defined for more than two random variables. For example,

$$F(x, y, z) = P(X \leq x, Y \leq y, Z \leq z)$$

is the joint distribution function of random variables X, Y, and Z, and they are independent if $F(x, y, z) = F_X(x)F_Y(y)F_Z(z)$.

A collection of random variables are called independent, if any finitely many of them, say $X_1, X_2, \ldots, X_n$, are independent, that is, if their joint distribution function F is given by

$$F(x_1, x_2, \ldots, x_n) = F_{X_1}(x_1)F_{X_2}(x_2) \cdots F_{X_n}(x_n).$$

Indicator: For an event A, the random variable defined by

$$1_A = \begin{cases} 1, & \text{on } A \\ 0, & \text{on } A^c \text{ (complement of } A), \end{cases}$$

is called the indicator of A. A collection of events are independent as defined before if their indicators are independent. A collection of events and random variables are called independent if they are independent when the events are identified with their indicators.

Convolution: The convolution of two distribution functions F and G is defined by

$$(F * G)(x) = \int_{-\infty}^{\infty} F(x - y)dG(y).$$

Then $F * G$ is also a distribution (function). The convolution is commutative: $F * G = G * F$, and is associative: $(F * G) * H = F * (G * H)$ for any three distributions. Thus, the convolution of any number of distributions, $F_1 * F_2 * \cdots * F_n$, is well defined. When all F_i are the same, we obtain an n-fold convolution, which is denoted as F^{*n}.

Sums of independent random variables: The distribution of a sum of independent random variables is given by a convolution as stated in the following proposition.

Proposition 1.1 *If X_i for $1 \leq i \leq n$ are independent random variables with distributions F_i and $Y = \sum_{i=1}^{n} X_i$, then*

$$F_Y = F_1 * F_2 * \cdots * F_n.$$

Proof Let $Z = X + Y$ with X and Y independent. Then

$$F_Z(z) = P(X + Y \leq z) = \int \int_{x+y \leq z} dF_X(x) dF_Y(y)$$

$$= \int_{-\infty}^{\infty} F_X(z - y) dF_Y(y).$$

This proves for $n = 2$. To prove for $n = 3$, let $W = X + Y + Z$ be an independent sum. Then by the result for $n = 2$ and the associativity of convolution, $F_W = (F_{X+Y}) * F_Z = F_X * F_Y * F_Z$. The general n is proved by an induction. $\diamondsuit$

The convolution of pmf's or pdf's:

$$f * g(x) = \begin{cases} \sum_{y+z=x} f(y)g(z) = \sum_y f(x-y)g(y), & \text{(for pmf's } f \text{ and } g) \\ \int_{-\infty}^{\infty} f(x-y)g(y)dy, & \text{(for pdf's } f \text{ and } g). \end{cases}$$

Then $f * g = g * f$ and $(f * g) * h = f * (g * h)$.

It is easy to show that if X_i are independent discrete (resp. continuous) random variables with pmf's (resp. pdf's) f_i, and $Y = \sum_{i=1}^n X_i$, then Y is discrete (resp. continuous) with pmf (resp. pdf)

$$f_Y = f_1 * f_2 * \cdots * f_n.$$

1.5 Convergence of random variables

Convergence in distribution: A sequence of random variables X_n are said to converge to a random variable X in distribution if for any bounded continuous function f, $E[f(X_n)] \to E[f(X)]$ as $n \to \infty$. Let F_n and F be, respectively, the distribution functions of X_n and X. The convergence in distribution is denoted as $X_n \overset{d}{\to} X$, $X_n \overset{d}{\to} F$ or $F_n \overset{d}{\to} F$, and is equivalent to either of the following two conditions:

(a) $F_n(x) \to F(x)$ as $n \to \infty$ whenever F is continuous at x.

(b) $P(a \leq X_n \leq b) \to P(a \leq X \leq b)$ as $n \to \infty$ whenever $P(X = a) = P(X = b) = 0$.

In the literature, the convergence in distribution for more general types of random variables is defined, in particular, it is defined for multi-dimensional random variables, but this will not be discussed here. However, note that if X_n and X take values in a space of isolated points (that is, no point is a limit of other points, such as integers), then $X_n \overset{d}{\to} X$ is equivalent to $P(X_n = x) \to P(X = x)$ for all x.

Almost sure convergence: A sequence of random variables X_n are said to converge almost surely (a.s.) to a random variable X if $X_n \to X$ on Ω except on a null event, that is, an event of zero probability. This is also denoted as $X_n \to X$ a.s. or $P(X_n \to X) = 1$. It can be shown that $X_n \to X$ a.s. implies $X_n \overset{d}{\to} X$.

SLLN (Strong law of large numbers): Let X_n be an iid (independent and identically distributed) sequence of random variables with a common finite mean μ. Then

$$\frac{1}{n} \sum_{i=1}^{n} X_i \to \mu \quad a.s. \qquad \text{as } n \to \infty.$$

This holds even for $\mu = \infty$ when all $X_n \geq 0$.

Convergence of expectation: In general, the almost sure convergence $X_n \to X$ a.s. does not necessarily imply the convergence of expectations $E(X_n) \to E(X)$. However, as the expectation is just an integral on the probability space, the following results hold.

Dominated convergence: $E(X_n) \to E(X)$ if $X_n \to X$ a.s. and $|X_n| \leq Y$ with $E(Y) < \infty$. In particular, $E(X_n) \to E(X)$ if $|X_n|$ are bounded by a finite constant, which is called the bounded convergence.

Monotone convergence: If $X_n \geq 0$ and $X_n \uparrow X$ a.s., then $E(X_n) \uparrow E(X)$ even when $E(X) = \infty$. Here $X_n \uparrow X$ means $X_1 \leq X_2 \leq \cdots \leq X_n \to X$ as $n \to \infty$.

1.6   Laplace transform and generating functions

Laplace transform: The Laplace transform of a nonnegative random variable X with distribution function F is defined by

$$L(s) = E[e^{-sX}] = \int_{[0, \infty)} e^{-sx} dF(x), \qquad s \geq 0. \tag{1.17}$$

This may be denoted as $L_X(s)$ to indicate its dependence on X.

The Laplace transform of a distribution type function $F(x)$ is defined by (1.17) and is denoted as $\hat{F}(s)$. The Laplace transform of a density type function $f(x)$ is defined by $\hat{f}(s) = \int_0^\infty e^{-sx} f(x) dx$. Note that $\hat{F}(s) = \hat{f}(s)$ if $f(x)$ is the pdf of $F(x)$.

Properties of Laplace transform: (a) $L_{aX+b}(s) = e^{-bs} L_X(as)$;
(b) If X and Y are independent, then $L_{X+Y}(s) = L_X(s)L_Y(s)$.
(c) $X \stackrel{d}{=} Y$ iff (if and only if) $L_X(s) = L_Y(s)$ for $s \geq 0$.
(d) $E(X) = -L'(0+)$ and $\mathrm{Var}(X) = L''(0+)^2 - L'(0+)^2$.

Note on Laplace transform: For a nonnegative random variable X, its Laplace transform $L(s)$ is finite for $s \geq 0$, but it may be ∞ for $s < 0$. For a general random variable X, $L(s)$ may or may not be finite for $s \neq 0$. It is always finite at $s = 0$ with $L(0) = 1$. It can be shown that if $L(s)$ is finite at some $s_1 \neq 0$, then it is finite and is continuous on the whole interval between 0 and s_1. Moreover, it is in fact smooth, that is, has a continuous derivative of any order, inside the interval. In this case, $L(s)$ determines the distribution of X.

For the proofs of these properties, the reader is referred to [2], where the corresponding results for the moment generating function $E(e^{sX}) = L(-s)$ are established.

Probability generating functions: For a discrete random variable X taking only nonnegative integer values, it is often convenient to work with its pgf (probability generating function) $P(s)$ or $P_X(s)$ defined by

$$P_X(s) = \sum_{n=0}^{\infty} p_n s^n \quad \text{for } 0 \leq s \leq 1, \tag{1.18}$$

where $p_n = P(X = n)$. Note that

$$P_X(s) = E(s^X) = L_X(-\ln s) \quad \text{for } 0 < s \leq 1,$$

and $P_X(s)$ is continuous for $s \in [0, 1]$ and differentiable for $s \in (0, 1)$.

Properties of pgf: (a) $P_X(0) = p_0$, $P_X'(0+) = p_1$, and more generally, $P_X^{(n)}(0+) = n!p_n$ for $n \geq 0$.
(b) $P_X(1) = 1$ and $P_X'(1-) = \mu_X$ ($\leq \infty$).
(c) If X and Y are independent, then $P_{X+Y}(s) = P_X(s)P_Y(s)$.
(d) $X \stackrel{d}{=} Y$ iff $P_X(s) = P_Y(s)$ for $0 \leq s \leq 1$.

Pgf of a deficient random variable: Let $X \geq 0$ be an integer-valued random variable which takes value ∞. Its pgf $P_X(s)$ is still defined by (1.18) but now $\sum_{n=1}^{\infty} p_n < 1$. Then

$$P_X(s) = E(s^X; X < \infty) \quad \text{for } 0 < s \leq 1,$$

where

$$E(X; A) = E(X 1_A) = \int_A X dP$$

for any random variable X and event A. Some properties stated above should be modified. For example,

$$P_X(1) = P(X < \infty) \quad \text{and} \quad P'_X(1-) = E(X; X < \infty).$$

1.7 Examples of discrete distributions

Bernoulli distribution $B(p)$ $(0 < p < 1)$: $X = 0$ or 1 (success). Its pmf p_X, mean μ, variance σ^2, and pgf P_X are given below:

$$p_X(0) = q = 1-p \text{ and } p_X(1) = p, \quad \mu = p, \quad \sigma^2 = pq, \quad P_X(s) = q + ps.$$

Binomial distribution $B(n, p)$: $X = X_1 + \cdots + X_n$, where X_i are iid $B(p)$. It is the # of successes in n independent trials with $p = P(\text{success})$.

$$p_X(k) = \binom{n}{k} p^k q^{n-k} \quad (q = 1 - p) \text{ for } k = 0, 1, 2, \ldots, n,$$
$$\mu = np, \quad \sigma^2 = npq, \quad P_X(s) = (q + ps)^n,$$

where $\binom{n}{k} = n!/[k!(n-k)!]$ is the # of ways to choose k objects from a group of n.

Geometric distribution $G(p)$ $(0 < p < 1)$: # of trials until the first success,

$$p_X(k) = q^{k-1} p \text{ for } k = 1, 2, \ldots, \quad \mu = \frac{1}{p}, \quad \sigma^2 = \frac{q}{p^2}, \quad P_X(s) = \frac{ps}{1 - qs}.$$

Poisson distribution Poisson(λ) $(0 < \lambda < \infty)$: This is the limit in distribution of $B(n, p)$ as $n \to \infty$, $p \to 0$ and $np \to \lambda$.

$$p_X(k) = e^{-\lambda}\frac{\lambda^k}{k!} \quad \text{for } k = 0, 1, 2, \ldots, \qquad \mu = \sigma^2 = \lambda, \qquad P_X(s) = e^{-\lambda+\lambda s}.$$

Law of rare events: For each integer $n \geq 1$, let X_{nj} for $j = 1, 2, 3 \ldots, n$ be independent random variables taking only values 0 and 1, and let $p_{nj} = P(X_{nj} = 1)$. If $\max_{1 \leq j \leq n} p_{nj} \to 0$ and $\sum_{j=1}^{n} p_{nj} \to \lambda$ as $n \to \infty$ with $0 < \lambda < \infty$, then $\sum_{j=1}^{n} X_{nj} \xrightarrow{d}$ Poisson(λ).

For the proof, see [9, Proposition 1.5.2].

1.8 Examples of continuous distributions

Uniform distribution $U(a, b)$: Its pdf $f(x)$, distribution function $F(x)$, mean μ, variance σ^2, and Laplace transform $L(s)$ are given below:

$$f(x) = 1/(b-a) \quad \text{for } a < x < b \quad (\text{and } f(x) = 0 \text{ otherwise}),$$

$$F(x) = \begin{cases} 0, & \text{for } x \leq a \\ (x-a)/(b-a), & \text{for } a \leq x \leq b \\ 1, & \text{for } x \geq b, \end{cases}$$

$$\mu = \frac{a+b}{2}, \quad \sigma^2 = \frac{(b-a)^2}{12}, \quad L(s) = \frac{e^{-as} - e^{-bs}}{s(b-a)} \quad (s \neq 0).$$

In particular, for $U(0, 1)$, $f(x) = 1$ and $F(x) = x$ for $0 \leq x \leq 1$, $\mu = 1/2$, $\sigma^2 = 1/12$, and $L(s) = (1 - e^{-s})/s$ $(s \neq 0)$.

Exponential distribution Exp(λ): $f(t) = \lambda e^{-\lambda t}$ for $t > 0$.

$$F(t) = 1 - e^{-\lambda t} \quad \text{for } t \geq 0, \qquad \mu = \frac{1}{\lambda}, \qquad \sigma^2 = \frac{1}{\lambda^2}, \qquad L(s) = \frac{\lambda}{\lambda + s}.$$

The parameter λ is the reciprocal of the mean, and is called the rate of the exponential distribution, because given any iid sequence of random variables X_n of mean $1/\lambda$, regarded as times between successive events, the number of events by time t is approximately λt for large t. Indeed, by the SLLN,

$$\lim_{t \to \infty} \frac{\# \text{ of events by time } t}{t} = \lim_{n \to \infty} \frac{n}{X_1 + \cdots + X_n} = \frac{1}{E(X_1)} = \lambda.$$

Lack of memory property of exponential distribution:

$$P(X > t + s \mid X > t) = P(X > s) \quad \text{for } t > 0 \text{ and } s > 0.$$

A random variable $X \geq 0$ is exponential iff it has the lack of memory property.

Minimum of independent exponential random variables:

Let $X_1, X_2, \ldots, X_n$ be independent exponential random variables of rates $\lambda_1, \lambda_2, \ldots, \lambda_n$ respectively. Then $X = \min(X_1, X_2, \ldots, X_n)$ is exponential of rate $(\lambda_1 + \cdots + \lambda_n)$.

Erlang distribution Erlang(n, λ):

Sum of n independent $\text{Exp}(\lambda)$.

$$f(t) = \frac{\lambda^n t^{n-1}}{(n-1)!} e^{-\lambda t} \text{ for } t > 0, \quad \mu = \frac{n}{\lambda}, \quad \sigma^2 = \frac{n}{\lambda^2}, \quad L(s) = \frac{\lambda^n}{(\lambda + s)^n}.$$

Normal distribution $N(\mu, \sigma^2)$:

$$f(x) = \frac{1}{\sqrt{2\pi}\,\sigma} \exp(-\frac{(x - \mu)^2}{\sigma^2}) \quad \text{for } -\infty < x < \infty,$$

$$L(s) = \exp(-\mu s + \frac{1}{2}\sigma^2 s^2) \quad \text{for } -\infty < s < \infty.$$

In particular, for standard normal $N(0, 1)$, $f(x) = \exp(-x^2/2)/\sqrt{2\pi}$ and $L(s) = \exp(s^2/2)$.

The table of standard normal probabilities is available in virtually every elementary probability or statistics text. For a general normal random variable X of mean μ and variance σ^2, $Z = (X - \mu)/\sigma$ is $N(0, 1)$, so the computation of the probability of X is reduced to a standard normal probability.

Central limit theorem (CLT):

Let X_n be iid with finite mean μ and variance σ^2, and let $\overline{X} = (X_1 + \cdots + X_n)/n$. Then

$$\frac{(X_1 + \cdots + X_n) - n\mu}{\sigma\sqrt{n}} = \frac{\overline{X} - \mu}{\sigma/\sqrt{n}} \xrightarrow{d} N(0, 1) \quad \text{as } n \to \infty.$$

Because of thes CLT, one often uses a normal distribution to approximately compute the probability of a large sum of iid random variables of an arbitrary or even unknown distribution as long as its mean and variance are known.

Simulation of a continuous random variable: Let U be $U(0,1)$ and let $F(x)$ be an arbitrary continuous distribution (function) with an inverse function F^{-1}. Then

$$X = F^{-1}(U)$$

is a continuous random variable with distribution F.

Thus, in order to simulate a continuous random variable X with distribution $F(x)$, one just needs to generate a random number U, uniformly distributed on $(0, 1)$, and set $X = F^{-1}(U)$. The MATLAB® command for generating U is rand.

For $\text{Exp}(\lambda)$, $F(x) = 1 - e^{-\lambda x}$ for $x > 0$, then $F^{-1}(y) = (1/\lambda)\ln(1 - y)$ and $X = (1/\lambda)\ln(1 - U)$. Since $1 - U$ is also $U(0,1)$, a simple way to simulate $\text{Exp}(\lambda)$ is given by

$$X = \frac{1}{\lambda}\ln U.$$

For example, the MATLAB® command for generating $\text{Exp}(2)$ is $(1/2)*\log(\text{rand})$.

1.9 Stochastic processes

Definition and classification: A stochastic process, or a process for short, is just a family of random variables indexed by time. The time index may be either discrete, such as $n = 0, 1, 2, \ldots$, or continuous, such as $t \in \mathbb{R}_+ = [0, \infty)$. The possible values of these random variables are called states of the process. The set of all states is called the state space of the process, which may also be either discrete, such as the set $\mathbb{Z}$ of integers, or continuous, such as the set $\mathbb{R}$ of real numbers. Thus, stochastic processes are sometimes classified into four classes: discrete time and discrete state, discrete time and continuous state, continuous time and discrete state, and continuous time and continuous state.

For example, a die is rolled successively and let X_n for $n = 1, 2, \ldots$ be the outcomes of successive rolls. This is a discrete time and discrete state process. A store manager may be interested in monitoring the customers entering the store. Let N_t be the number of customers entering during the time interval $[0, t]$. This is a continuous time and discrete state process. The amount X_t of sales during the time interval $[0,\ t]$ is a continuous time and (possibly) continuous state process.

For typographical convenience, X_t may be written as $X(t)$.

Distribution of a process: For a process $X(t)$, the distribution of the random variable $X(t)$ for a fixed time t is called a 1-dimensional marginal distribution of the process. The joint distribution of $X(t_1), X(t_2), \ldots, X(t_n)$, for finitely many time points $t_1 < t_2 < \cdots < t_n$, is called a finite dimensional distribution of the process $X(t)$. The distribution of process X_t is referred to all its finite dimensional distributions. Thus, two processes X_t and Y_t are said to have the same distribution if all the finite dimensional distributions are the same. When two processes have the same distribution, then any probability computation done for one process will be identical to that for the other. However, processes with the same distribution may have quite different pathwise properties.

Two processes may have the same 1-dimensional marginal distributions, but not the same distribution. For example, when a fair die is rolled successively, let X_n be the outcomes of successive rolls and let $Y_n = X_1$ for all $n \geq 1$. Then $X_n \overset{d}{=} Y_n$ for each n, but the two processes do not have the same distribution as $P(X_1 = X_2 = 6) = 1/36 \neq 1/6 = P(Y_1 = Y_2 = 6)$.

Existence of processes: The finite dimensional distributions of a process $X(t)$ are consistent in the sense that for any two sets of time points, where one is contained in the other, the joint distribution over the smaller set is the restriction of that over the larger set. For example,

$$P[X(t_1) \in B_1, X(t_2) \in B_2, X(t_3) \in \mathbb{R}] = P[X(t_1) \in B_1, X(t_2) \in B_2]$$

for any time points t_1, t_2, t_3, and Borel subsets B_1 and B_2 of $\mathbb{R}$. In fact, before there is a process, one may specify its finite dimensional distributions; as long as they are consistent, the process exists as guaranteed by Kolmogorov's Theorem for the existence of processes (see [2]).

Convergence in distribution: A sequence of processes $X^n(t)$ are said to converge in distribution to a process $X(t)$ if any finite dimensional distribution of $X^n(t)$ converges to that of $X(t)$, that is, if for any $t_1 < t_2 < \cdots < t_n$,

$$(X^n(t_1), X^n(t_2), \ldots, X^n(t_n)) \overset{d}{\to} (X(t_1), X(t_2), \ldots, X(t_n)).$$

In the literature, the convergence in distribution of processes may have a stronger definition.

Path regularity: A process X_t is called right continuous if all its paths are right continuous in t, that is, if $X_t = X_{t+}$ for all t. It is called continuous if all its paths are continuous in t. It is called increasing if all its paths are increasing in t, that is, if $X_s \leq X_t$ for any $s < t$.

It may be possible to modify the process X_t on a null event for each fixed t so that the modified process will possess a certain path regularity. It is clear that the modification will not change the distribution of the process.

1.10 Stopping times

Definition: A stopping time τ of a discrete time process X_n is a nonnegative integer-valued random variable, possibly taking value ∞, such that for any $n \geq 0$, the event $[\tau = n]$ is completely determined by the process up to time n. This means that to determine whether $\tau = n$, one just needs to look at the values of $X_1, X_2, \ldots, X_n$. For example, let X_n be the successive outcomes of tossing a coin, then the first time τ when heads appear is a stopping time of the process X_n. The same is true for the second or the nth time when heads appear, but the last time before some fixed time when this happens is not a stopping time.

Note that

$$[\tau \leq n] = \cup_{i=0}^{n}[\tau = i] \quad \text{and} \quad [\tau = n] = [\tau \leq n] \cap [\tau \leq n-1]^c.$$

Therefore, in the above definition of stopping time, one may replace $[\tau = n]$ by $[\tau \leq n]$. The stopping time of a continuous time process X_t is defined to be a nonnegative-valued random variable τ, possibly taking value ∞, such that for any $t \geq 0$, the event $[\tau \leq t]$ is determined by the process up to time t. One cannot replace $[\tau \leq t]$ by $[\tau = t]$ in the definition of stopping times for a continuous time process.

The informal statement that an event A, or a random variable Z, is determined by a collection of random variables $X, Y, \ldots$ can be made more precise. Formally, this means that A is contained in, or Z is measurable under, the σ-algebra $\sigma\{X, Y, \ldots\}$ generated by $X, Y \ldots$, which by definition is the smallest σ-algebra containing all the events $[X \leq x], [Y \leq y], \ldots$ for any real $x, y, \ldots$.

An extended notation of stopping time: Two families of random variables $\{X_1, X_2, \ldots\}$ and $\{Y_1, Y_2, \ldots\}$ are called independent if any event in $\sigma\{X_1, X_2, \ldots\}$ is independent of any event in $\sigma\{Y_1, Y_2, \ldots\}$. In this sense, one may talk about the independence between two processes, and in particular, between a random variable and a process.

It is often useful to slightly extend the definition of a stopping time τ of a process X_n given above by allowing $[\tau \leq n]$ to be determined not only by X_k for $k \leq n$, but also by some random variables independent of the process X_n. In particular, any nonnegative random variable independent of a process is a stopping time of the process. In the sequel, a stopping time is understood in this extended sense unless when explicitly stated otherwise.

Theorem 1.2 *(Wald's identity) Let X_n be iid with $E(X) < \infty$, where X represents a random variable with same distribution as X_n (it is a common practice to omit the index of a generic term of an iid sequence), and let $\tau \geq 0$ be a stopping time of $\{X_n\}$ with $E(\tau) < \infty$. Then*

$$E(\sum_{n=1}^{\tau} X_n) = E(\tau)E(X).$$

Moreover, if $X_n \geq 0$, then the above holds even when either $E(X)$ or $E(\tau)$ is infinite.

Proof: Recall 1_A is the indicator function of an event A.

$$E(\sum_{n=1}^{\tau} X_n) = E[\sum_{n=1}^{\infty} X_n 1_{[\tau \geq n]}] = \sum_{n=1}^{\infty} E[X_n 1_{[\tau \geq n]}]$$

$$= \sum_{n=1}^{\infty} E(X_n)P(\tau \geq n)$$

(because $[\tau \geq n] = [\tau \leq n-1]^c$ is independent of X_n)

$$= E(X) \sum_{n=1}^{\infty} P(\tau \geq n) = E(X)E(\tau). \quad \diamond$$

Example 1.3 Consider a fair game of gambling in which the gambler enters a series bets of \$1 each and the probability of winning a bet is $1/2$. If the gambler stops after n bets, his expected net gain is 0 because of the fair probability $1/2$. The question is whether he can expect a positive net gain by stopping at a random time. By Wald's identity, if he stops at a stopping time with a finite mean, such as at the time of tenth win, for example, then the expected net gain is still 0. What about stopping at the first time τ when his net gain is positive? It is clear that τ is a stopping time and his net gain is positive at time τ. The apparent contradiction to Wald's identity implies that $E(\tau) = \infty$.

Exercise 1.1 Successive insurance claims are iid exponential of mean 500. Find the expected total accumulated claim at the time when a claim exceeds 2000.

Exercise 1.2 Let X_n be iid with $P(X = 1) = p$ and $P(X = -1) = 1 - p$ for $0 < p < 1$, let $Y_n = \sum_{i=1}^{n} X_i$, and let τ be the first time when $Y_n = m$ for some integer $m > 0$. Use Wald's identity to find $E(\tau)$ and determine when $E(\tau) = \infty$.

Exercise 1.3 A miner is trapped in a room with four doors. The first three doors lead him back to the same room after 1, 2, and 3 hours of journey, respectively, and then he has to try again, but the fourth door leads him to safety after 4 hours. Suppose all four doors have equal chance to be chosen. Find the expected time when the miner reaches safety.
Note: Use Walt's identity to solve the problem assuming the miner does not remember the door he has chosen before. The problem, of course, will have a different solution if he remembers the door; see Exercise 1.4.

1.11 Conditional expectation

Conditional expectation given an event: Let A be an event with $P(A) > 0$. Recall $E(X; A) = E(X1_A)$ for a random variable X. Assume $E(X)$ is finite. Define the conditional expectation of X given A by

$$E(X \mid A) = E(X; A)/P(A).$$

This is just the expectation of X under the conditional probability P_A given A.

Conditional expectation given a discrete random variable:
Let X be a general random variable with finite $E(X)$ and let Y be a discrete random variable. The conditional expectation of X given Y is a random variable, denoted as $E(X \mid Y)$, defined by

$$E(X \mid Y) = E(X \mid Y = y) \quad \text{on the event } [Y = y] \tag{1.19}$$

for y varying over all possible values of Y.

The random variable $Z = E(X \mid Y)$ is a function of Y, that is, $Z = f(Y)$ for some Borel function $f(y)$. Indeed, $f(y) = E(X \mid Y = y)$ if y is a possible value of Y and $f(y)$ may be arbitrarily defined, say $f(y) = 0$, if y is not a possible value of Y. Although the choice for $f(y)$ is not unique, $Z = f(Y)$ is uniquely determined up to a null event. Moreover, for any event A in the σ-algebra $\sigma\{Y\}$ generated by Y (defined in §1.10),

$$E(Z; A) = E(X; A). \tag{1.20}$$

This may be easily verified as any $A \in \sigma\{Y\}$ is a union of events of the form $[Y = y]$. On the other hand, it is easy to show that if Z is a random variable that is a function of Y, and satisfies (1.20) for any $A \in \sigma\{Y\}$, then $Z = E(X \mid Y)$ a.s.

Conditional expectation given a general random variable: For any two random variables X and Y with finite $E(X)$, it can be shown that there is a random variable Z, unique almost surely, that is a function of Y and satisfies (1.20) for any $A \in \sigma\{Y\}$. We define Z to be the conditional expectation of X given Y and denote it by $E(X \mid Y)$.

In general, $P(Y = y)$ may be 0, and so $E(X \mid Y = y)$ may not be defined in the usual sense, but as $E(X \mid Y)$ is a function of Y, say $f(Y)$, we may define

$$E(X \mid Y = y) = f(y)$$

even when $P(Y = y) = 0$.

Jointly continuous random variables: Let $f(x, y)$ be the joint pdf of X and Y. Then

$$E(X \mid Y) = \frac{\int_{-\infty}^{\infty} x f(x, Y) dx}{f_Y(Y)}, \tag{1.21}$$

where $f_Y(y) = \int_{-\infty}^{\infty} f(x, y) dx$ is the pdf of Y. Because $P[f_Y(Y) > 0] = 1$, so the expression in (1.21) is a.s. well defined.

To prove (1.21), one needs to show (1.20) holds for $Z = E(X \mid Y)$ given in (1.21). An easy computation using the joint pdf $f(x, y)$ shows that $E[Zh(Y)] = E[Xh(Y))]$ for any bounded Borel function $h(y)$, which implies (1.20).

Conditional expectation as a limit: Let X and Y be two general random variables with X being bounded. For $\varepsilon > 0$, let $I_\varepsilon(y)$ be an interval of length ε containing y, such as $(y - \varepsilon/2, y + \varepsilon/2)$ or $[y, y + \varepsilon)$. Then

$$E(X \mid Y = y) = \lim_{\varepsilon \to 0} P[X \mid Y \in I_\varepsilon(y)]. \qquad (1.22)$$

More precisely, let Λ be the set of all real numbers y such that $P[Y \in I_\varepsilon(y)] > 0$ for any $I_\varepsilon(y)$ ($\varepsilon > 0$). Then (1.22) holds for $y \in \Lambda$, assuming the limit exists for all $y \in \Lambda$ along any sequence of intervals $I_\varepsilon(y)$ with $\varepsilon \to 0$. Because $P[Y \in \Lambda^c] = 0$, $E(X \mid Y = y)$ may be defined arbitrarily for $y \in \Lambda^c$.

In an actual computation of $E(X \mid Y = y)$ in (1.22), we may take a special form of interval $I_\varepsilon(y)$, say $I_\varepsilon(y) = [y, y + \varepsilon)$, with the understanding that the limit is assumed to exist and be the same along any sequence of intervals containing y and of length ε, as $\varepsilon \to 0$.

Proof of (1.22): Let $g(y)$ be the limit in (1.22), we want to show $E[Xh(Y)] = E[g(Y)h(Y)]$ for any bounded Borel function h. By a standard argument in the measure theory, it suffices to prove this for a bounded continuous h. Then

$$E[Xh(Y)] = \sum_k E[Xh(Y); \frac{k}{n} \le Y < \frac{k+1}{n}]$$

$$= \lim_{n \to \infty} \sum_k E[X; \frac{k}{n} \le Y < \frac{k+1}{n}]h(\frac{k}{n})$$

$$= \lim_{n \to \infty} \sum_k E[X \mid \frac{k}{n} \le Y < \frac{k+1}{n}]h(\frac{k}{n})P[\frac{k}{n} \le Y < \frac{k+1}{n}]$$

$$= \lim_{n \to \infty} E[g_n(Y)h_n(Y)],$$

where $g_n(y) = P[X \mid \frac{k}{n} \le Y < \frac{k+1}{n}]$ and $h_n(y) = h(\frac{k}{n})$ for $y \in [\frac{k}{n}, \frac{k+1}{n})$. Because $g_n(y) \to g(y)$ and $h_n(y) \to h(y)$ as $n \to \infty$, by the bounded convergence, $E[Xh(Y)] = \lim_{n \to \infty} E[g_n(Y)h_n(Y)] = E[g(Y)h(Y)]$. ◇

Properties: Assume $E(X)$, $E(X_1)$ and $E(X_2)$ below are finite.

(a) $E(a_1 X_1 + a_2 X_2 \mid Y) = a_1 E(X_1 \mid Y) + a_2 E(X_2 \mid Y)$ for any constants a_1 and a_2.

(b) $E(X \mid Y) = X$ if X is a function of Y.

(c) $E(X \mid Y) = E(X)$ if X is independent of Y.

(d) Total probability law for expectation:

$$E(X) = E[E(X \mid Y)] \;=\; \int_{-\infty}^{\infty} E(X \mid Y = y) dF_Y(y)$$

$$= \begin{cases} \sum_y E(X \mid Y = y) p_Y(y) & \text{(disc. } Y) \\ \int_{-\infty}^{\infty} E(X \mid Y = y) f_Y(y) dy & \text{(cont. } Y). \end{cases}$$

These properties can be easily verified for a discrete random variable Y but hold in general. Note that any equality involving conditional expectation given a random variable should be understood to hold almost surely as conditional expectations are so defined.

Note that $E(X \mid Y)$ is defined even when $E(X) = \infty$ provided $X \geq 0$. In this case, properties (b), (c), and (d) still hold, and (a) also holds when $X_1, X_2, a_1, a_2 \geq 0$.

Conditional probability given a random variable Y:

$$P(A \mid Y) = E(1_A \mid Y).$$

The total probability law of an event A given a random variable Y is

$$P(A) = E[P(A \mid Y)] = \int_{-\infty}^{\infty} P(A \mid Y = y) dF_Y(y). \qquad (1.23)$$

Conditional expectation given two or more random variables:
Assume $E(X)$ is finite or $X \geq 0$. The conditional expectation of X given two discrete random variables Y and Z is defined by

$$E(X \mid Y, Z) = E(X \mid Y = y, Z = z) \quad \text{on } [Y = y, Z = z].$$

This is a function of Y and Z such that for any $A \in \sigma\{Y, Z\}$,

$$E[E(X \mid Y, Z); A] = E(X; A). \qquad (1.24)$$

In general, for any three random variables X, Y, Z, assuming $E(X)$ is finite or $X \geq 0$, there is a function of Y and Z, denoted as $E(X \mid Y, Z)$ and called the conditional expectation of X given Y and Z, such that

(1.24) holds for any $A \in \sigma\{Y, Z\}$. The total probability law now takes the form:

$$E(X) = E[E(X \mid Y, Z)]. \tag{1.25}$$

The conditional expectation given more than two random variables is defined similarly. All the properties of the conditional expectation given a single random variable, stated earlier, hold also when given two or more random variables.

Example 1.4 Let X and Y be two independent exponential random variables of rates λ_1 and λ_2, respectively. Find $P(X > Y)$.

Solution: $P(X > Y)$ may be obtained directly by computing a double integral of the joint pdf $f(x, y) = \lambda_1 e^{-\lambda_1 x} \lambda_2 e^{-\lambda_2 y}$ of X and Y,

$$P(X > Y) = \int_0^\infty \int_0^x \lambda_1 e^{-\lambda_1 x} \lambda_2 e^{-\lambda_2 y} \, dy \, dx.$$

Alternatively, it may be obtained using the conditional expectation as follows.

$$P(X > Y) = E[P(X > Y \mid Y)] = \int_{-\infty}^\infty P(X > Y \mid Y = y) dF_Y(y)$$

$$= \int_{-\infty}^\infty P(X > y \mid Y = y) dF_Y(y) = \int_0^\infty P(X > y) dF_Y(y)$$

(because X and Y are independent)

$$= \int_0^\infty e^{-\lambda_1 y} dF_Y(y) = L_Y(\lambda_1) = \lambda_2/(\lambda_1 + \lambda_2).$$

In short hand, $P(X > Y) = E[P(X > Y \mid Y)] = E[e^{-\lambda_1 Y}] = L_Y(\lambda_1) = \lambda_2/(\lambda_1 + \lambda_2)$.

Example 1.5 Let X and Y be two independent random variables, and one of them is continuous. Show that $P(X = Y) = 0$.

Solution: Suppose X is continuous. Then

$$P(X = Y) = E[P(X = Y \mid Y)]$$

$$= \int_{-\infty}^\infty E(X = y \mid Y = y) dF_Y(y) = \int_{-\infty}^\infty P(X = y) dF_Y(y) = 0.$$

Example 1.6   A fair die is rolled to produce a number and a fair coin is tossed that number of times.
(a) Find the expected number of heads obtained.
(b) Find the probability of obtaining 4 heads.

Solution: (a) Let X be the number of heads obtained and let Y be the number on the die. Then $E(X) = E[E(X \mid Y)] = E[Y/2] = E(Y)/2 = (7/2)/2 = 7/4$.
(b) $P(X = 4) = E[P(X = 4 \mid Y)] = E[\binom{Y}{4}(1/2)^Y; Y \geq 4] = [\binom{4}{4}(1/2)^4 + \binom{5}{4}(1/2)^5 + \binom{6}{4}(1/2)^6](1/6) = 0.0755.$

Example 1.7   As in Exercise 1.2, let X_n be iid with $P(X = 1) = p$ and $P(X = -1) = 1 - p$ for $0 < p < 1$, and let $Y_n = X_1 + \cdots + X_n$ ($Y_0 = 0$). The discrete time process Y_n is called a simple random walk. Let τ be the first time n when $Y_n = m$ for some integer $m > 0$. Find the pgf $P_\tau(s)$ of τ, and use it to find $P(\tau < \infty)$ and $E(\tau)$ (compare with Exercise 1.2).

Solution: We will first solve for $m = 1$. For $s > 0$,

$$
\begin{aligned}
P_\tau(s) &= E(s^\tau; \tau < \infty) = E[E(s^\tau; \tau < \infty \mid X_1)] \\
&= E(s \mid X_1 = 1)P(X_1 = 1) + \\
&\quad E(s^{1+\tau_1+\tau_2}; \tau_1 < \infty, \tau_2 < \infty \mid X_1 = -1)P(X_1 = -1) \\
&\quad (\tau_1 \text{ is time from } -1 \text{ to } 0 \text{ and } \tau_2 \text{ is time from } 0 \text{ to } 1, \\
&\quad \text{they are independent with same distribution as } \tau) \\
&= sp + sP_\tau(s)^2(1 - p).
\end{aligned}
$$

Solve the quadratic equation for $P_\tau(s)$ to get

$$
P_\tau(s) = \frac{1 \pm \sqrt{1 - 4s^2 p(1 - p)}}{2s(1 - p)} = \frac{1 - \sqrt{1 - 4s^2 p(1 - p)}}{2s(1 - p)}.
$$

(The $+$ sign is dropped because it leads to $P_\tau(s) \to \infty$ as $s \to 0$). From this, and noting $1 - 4p(1 - p) = (1 - 2p)^2$, we see that

$$
\begin{aligned}
P(\tau < \infty) = P_\tau(1) &= \frac{1 - \sqrt{1 - 4p(1 - p)}}{2(1 - p)} \\
&= \frac{1 - |1 - 2p|}{2(1 - p)} = \begin{cases} 1, & \text{if } p \geq 1/2 \\ p/(1 - p), & \text{if } p < 1/2 \end{cases}
\end{aligned}
$$

For $p \geq 1/2$, differentiate $P_\tau(s)$ and then let $s \uparrow 1$; after some rather tedious computation, we get

$$E(\tau) = P'_\tau(1-) = \begin{cases} 1/(2p-1) & \text{if } p > 1/2 \\ \infty & \text{if } p = 1/2. \end{cases}$$

Of course $E(\tau) = \infty$ when $p < 1/2$ as $P(\tau < \infty) = p/(1-p) < 1$ in this case.

Now for $m > 1$, $\tau = \tau_1 + \tau_2 + \cdots + \tau_m$, where τ_1 is the time from 0 to 1, τ_2 is the time from 1 to 2, ..., τ_m is the time from $m-1$ to m. Then $\tau_1, \tau_2, \ldots, \tau_m$ are iid, and

$$P_\tau(s) \;=\; P_{\tau_1}(s)^m = [\frac{1 - \sqrt{1 - 4s^2 p(1-p)}}{2s(1-p)}]^m,$$

$$P(\tau < \infty) \;=\; \begin{cases} 1, & \text{if } p \geq 1/2 \\ p^m/(1-p)^m, & \text{if } p < 1/2, \end{cases}$$

$$E(\tau) \;=\; \begin{cases} m/(2p-1), & \text{if } p > 1/2 \\ \infty, & \text{if } p \leq 1/2. \end{cases}$$

Exercise 1.4 Re-solve the miner problem in Exercise 1.3 using the conditioning method of this section under each of the following assumptions.
(a) Assume the miner does not remember the doors used before, as in Exercise 1.3.
(b) Assume he remembers only the door tried the last time.
(c) Assume he remembers all the doors tried previously.

Exercise 1.5 Suppose X_n are iid exponential of rate λ, and N is geometric with $p = P(N = 1)$. Assume N and X_n are independent. Let $Y = X_1 + X_2 + \cdots + X_N$. Show Y is exponential of rate λp. Hint: Find the Laplace transform of Y by conditioning on N.

Exercise 1.6 A motorist will be stopped by police a Poisson number of times of mean 5 on a trip. Each time he is stopped by police, he will be fined 0, 100, and 200 with probabilities 0.2, 0.5, and 0.3, respectively. Find the probability that the total fine exceeds 200.
Hint: Let N be the number of police stops and let X_n be the nth fine. Then $Y = \sum_{n=1}^{N} X_n$ is the total fine. Let 1 represent $100. First find the pgf $P_Y(s)$ and use it to get $P(Y > 2)$.

Exercise 1.7 Three people A, B, and C enter a post office at the same time. A and B are immediately served by the two available servers, and C waits for his turn. Suppose the service times for A, B, and C are independent and exponential of means 3, 4, and 5, respectively. Find the probability that C is not the last to leave.

Exercise 1.8 Let X and Y be independent random variables of uniform distribution $U(0,1)$. Let

$$Z = \min(X,Y) \quad \text{and} \quad W = \max(X,Y).$$

(a) Find $E(Z \mid W)$.
(b) Find $E(X \mid W)$.
Hint: Use either (1.21) or (1.22) to solve (a), but only (1.22) to solve (b) because X and W are not jointly continuous.

Exercise 1.9 Consider a small extension of the simple random walk in Example 1.7 by allowing the random steps X_n to take value 0. Precisely, let X_n be iid with

$$p = P(X_1 = 1), \quad q = P(X_1 = -1), \quad r = P(X_1 = 0),$$

where $p > 0$, $q > 0$, $r \geq 0$ and $p + q + r = 1$. Let $Y_0 = 0$ and $Y_n = X_1 + \cdots + X_n$ for $n \geq 1$. Find the pgf $P_\tau(s)$ of the first time τ when $Y_n = 1$, $P(\tau < \infty)$ and $E(\tau)$.

Chapter 2

Poisson processes

2.1 Introduction to Poisson processes

Counting processes: A continuous time process $N(t)$ taking integer values is called a counting process if it is right continuous and increasing. It may be regarded as counting the number of certain events up to time t. A counting process $N(t)$ jumps up whenever an event occurs and the size of jump $N(t) - N(t-)$ is the number of events occurring at time t.

A counting process $N(t)$ is called simple if the sizes of jumps are all equal to 1. In this case, two events cannot occur at the same time.

Independent and stationary increments: Let $N(t)$ be a counting process. Then $N(s, t] = N(t) - N(s)$ is the number of events occurring in the time interval $(s, t]$ for $s < t$, called the increment of process over the interval $(s, t]$. The process $N(t)$ is said to have independent increments if the increments over non-overlapping time intervals are independent, that is, for any $0 < t_1 < t_2 < \cdots < t_n$,

$$N(0), \quad N(0, t_1], \quad N(t_1, t_2], \quad \ldots, \quad N(t_{n-1}, t_n]$$

are independent. This is equivalent to saying that for $s < t$, $N(s, t]$ is independent of the process up to time s, that is, independent of $\{N(u); u \leq s\}$.

The process $N(t)$ is said to have stationary increments if the distribution of $N(s, t]$ for any $s < t$ depends only on $t - s$. Then $N(s, t] \overset{d}{=} N(0, t - s] = N(t - s) - N(0)$.

Poisson processes: A counting process $N(t)$ with $N(0) = 0$ is called a Poisson process of rate λ, where $\lambda > 0$ is some constant, if

(a) $N(t)$ has independent and stationary increments; and
(b) $N(t)$ has Poisson distribution of mean λt.

Thus, if $N(t)$ is a Poisson process of rate λ, then $N(s, t]$ is Poisson of mean $\lambda(t - s)$.

Theorem 2.1   *A Poisson process is a simple counting process.*

Proof: For any $t > 0$, $P(\text{two events occur at same time in } [0, t])$

$$
\leq \sum_{j=1}^{n} P(\text{at least two events in } [\tfrac{(j-1)t}{n}, \tfrac{jt}{n}]) = \sum_{j=1}^{n} [1 - e^{-\lambda t/n}(1 + \frac{\lambda t}{n})
$$

$$
= \sum_{j=1}^{n} e^{-\lambda t/n}(e^{\lambda t/n} - 1 - \frac{\lambda t}{n}) = \sum_{j=1}^{n} e^{-\lambda t/n}(\frac{\lambda^2 t^2}{2! n^2} + \cdots) \leq \frac{\lambda^2 t^2}{2n} \to 0
$$

as $n \to \infty$. $\diamondsuit$

Theorem 2.2   *Let $N(t)$ be a simple counting process with $N(0) = 0$. Assume that it has independent and stationary increments, and $\lambda = E[N(1)]$ is > 0 and finite. Then $N(t)$ is a Poisson process of rate λ.*

Proof: We need to show that $N(t)$ is a Poisson of mean λt. Because of stationary increments, for integers $m, n > 0$, $E[N(1)] = nE[N(1/n)]$ and $E[N(m/n)] = mE[N(1/n)] = (m/n)E[N(1)] = \lambda(m/n)$. By the right continuity of $N(t)$, $E[N(t)] = \lambda t$ for any $t > 0$. It remains to show that $N(t)$ has a Poisson distribution.

Divide the time interval $(0, t]$ into n subintervals of equal length t/n and let X_n be the number of subintervals during which an event occurs. Then X_n is binomial $B(n, p)$ with $p = P[N(t/n) > 0]$. Because of right continuity of $N(t)$ and $N(0) = 0$, $p \to 0$ as $n \to \infty$. Because the counting process $N(t)$ is simple, $X_n \uparrow N(t)$ as $n \to \infty$, and hence, $np = E(X_n) \to E[N(t)] = \lambda t$. By the Poisson approximation of binomial distribution (see §1.7), X_n converges in distribution to a Poisson distribution of mean λt. $\diamondsuit$

Time homogeneity of Poisson process: Let $N(t)$ be a Poisson process. By the independent and stationary increments, for any constant $T > 0$, the process $N^T(t) = N(T + t) - N(t)$, called the process $N(t)$ time shifted by T, is a Poisson process of the same rate and is independent of the process $N(t)$ up to time T.

This fact is true even for a stopping time T. If $N(t)$ is a Poisson process of rate λ and T is a stopping time of $N(t)$, then given $T < \infty$,

$N^T(t)$ is a Poisson process of rate λ and is independent of process $N(t)$ up to time T. We will describe this more precisely below, but the reader may choose to accept the above intuitive description and go directly to Example 2.3.

More on time homogeneity: Let τ be a finite stopping time of a Poisson process $N(t)$ of rate λ, and let $N^\tau(t) = N(\tau+t) - N(\tau)$ be the process $N(t)$ time shifted by τ. Then $N^\tau(t)$ is a Poisson process of rate λ and is independent of the original Poisson process $N(t)$ up to time τ. This last statement means that the process $N(t)$ is independent of the stopped process $N^{\wedge\tau}(t) = N(t\wedge\tau)$, where $t\wedge\tau = \min(t, \tau)$. We will only outline the main ideas in the proof of this advanced topic. Fix an integer $n > 0$, define $\tau^n = (k + 1)/2^n$ on the event $[k/2^n \le \tau < (k + 1)/2^n]$ for $k = 0, 1, 2, 3, \ldots$. Then τ^n is a stopping time of process $N(t)$ and it takes only discrete values, and $\tau^n \downarrow \tau$. One may first prove for τ^n and then take the limit as $n \to \infty$.

More generally, for any stopping time τ with $P(\tau < \infty) > 0$, under the conditional probability given $[\tau < \infty]$, $N^\tau(t)$ is a Poisson process of the same rate as $N(t)$, and is independent of $N(t)$ up to τ.

Example 2.3 Suppose customers arrive in a store according to a Poisson process of rate $\lambda = 2$ customers per hour. The store opens at 8 a.m. Find
(a) $P(5$ customers arrive before 10 a.m. and total 6 arrivals by 11 a.m.);
(b) $P($less than 3 arrivals by 10 a.m. but at least 5 by 11 a.m.$)$;
(c) Suppose each customer has a probability of 0.4 to make a purchase. Find the probability that a purchase occurs by 10 a.m.

Solution: (a) Set 8 a.m. as time zero.

$$P[N(2) = 5, N(3) = 6] = P[N(2) = 5, N(2,3] = 1]$$
$$= P[N(2) = 5]P[N(2,3] = 1] = e^{-2(2)}\frac{[2(2)]^5}{5!} \cdot e^{-2(1)}[2(1)] = 0.0423.$$

(b)

$$P[N(2) < 3, \; N(3) \geq 5]$$
$$= \;\; P[N(2) = 0, \; N(2,3] \geq 5] + P[N(2) = 1, \; N(2,3] \geq 4]$$
$$+P[N(2) = 2, \; N(2,3] \geq 3]$$
$$= \;\; P[N(2) = 0]P[N(2,3] \geq 5] + P[N(2) = 1]P[N(2,3] \geq 4]$$
$$+P[N(2) = 2]P[N(2,3] \geq 3]$$
$$= \;\; e^{-4}\{1 - e^{-2}[1 + 2 + 2^2/2 + 2^3/3! + 2^4/4!]\}$$
$$+e^{-4}(4)\{1 - e^{-2}[1 + 2 + 2^2/2 + 2^3/3!]\}$$
$$+e^{-4}(4^2/2)\{1 - e^{-2}[1 + 2 + 2^2/2]\} \;\; = \;\; \ldots.$$

(c) If there are n arrivals by 10 a.m., then the probability of no purchase is $(1 - 0.4)^n = 0.6^n$. By the total probability law,

$$P(\text{no purchase by 10 a.m.}) = E\{P[\text{no purchase} \mid N(2)]\}$$
$$= \;\; E(0.6^{N(2)}) = P_{N(2)}(0.6) = e^{-2(2)+2(2)0.6} = 0.2019.$$

Then $P(\text{a purchase by 10 a.m.}) = 1 - 0.2019 = 0.7981$.

Exercise 2.1 Shocks occur to a system according to a Poisson process at an average rate of one every two hours. Find:
(a) $P(\text{at most 1 shock during the first hour and total of 3 shocks during the first four hours})$.
(b) Suppose each shock has a probability of 0.1 to destroy the system. Find the probability that the system is not destroyed after 3 hours.
(c) Suppose each shock causes damage to the system and requires a repair cost that is uniformly distributed between 0 and 2 if the system is not destroyed, and 10 if destroyed. Find the expected total repair cost after 3 hours.

Exercise 2.2 Customers arrive at a service station according to a Poisson process of rate $\lambda = 2$. The server may break down, independently of the arrivals, and each downtime lasts a random time uniformly on $(0, 1)$.
(a) Find the expected number of customers arriving in the next down period.
(b) Find the probability of two or more customers arriving in the next down period.

2.2 Arrival and inter-arrival times of Poisson processes

Arrival times: For a Poisson process $N(t)$ of rate λ, let S_n be the time when the nth event occurs, called the nth arrival time, and set $S_0 = 0$.

Example 2.4 Let $N(t)$ be a Poisson process of rate 2. Find
(a) $P(S_1 \leq 1, 1 < S_2 \leq 2)$;
(b) $P(S_1 \leq 1, S_2 \leq 2)$.

Solution: (a) $P(S_1 < 1, 1 < S_2 < 2) = P[N(1) = 1, N(1,2] \geq 1] = (e^{-2}2)(1 - e^{-2}) = 0.234$.
(b) $P(S_1 < 1, S_2 < 2) = P[N(1) = 1, N(1,2] \geq 1] + P[N(1) \geq 2]$
$= (e^{-2}2)(1 - e^{-2}) + (1 - e^{-2} - e^{-2}2) = 0.828$.

Inter-arrival times: Let $T_1 = S_1$, $T_2 = S_2 - S_1$, ..., $T_n = S_n - S_{n-1}$. Then T_n is the time between $(n-1)$st and nth arrivals, called the nth inter-arrival time.

Theorem 2.5 *For a Poisson process of rate λ, the inter-arrival times T_n are iid $Exp(\lambda)$. Consequently, S_n has Erlang distribution $Erlang(n, \lambda)$ for $n \geq 1$ with pdf $f(t) = \lambda^n t^{n-1} e^{-\lambda t}/(n-1)!$, $t > 0$.*

Proof: Because $P(T_1 > t) = P[N(t) = 0] = e^{-\lambda t}$, T_1 is $Exp(\lambda)$. Because T_2 is just T_1 for the time-shifted process $N^{S_1}(t) = N(S_1+t) - N(S_1)$, by the time homogeneity, T_2 is independent of $T_1 = S_1$ and has the same distribution. Similarly, shifting time to $S_2, S_3, \ldots$ proves that T_n are iid $Exp(\lambda)$. ◇

Distribution of a single arrival time: For $0 < a < b$,

$$P[a < S_n \leq b] = P[N(b) \geq n] - P[N(a) \geq n]$$
$$= e^{-\lambda a} \sum_{i=0}^{n-1} \frac{(\lambda a)^i}{i!} - e^{-\lambda b} \sum_{i=0}^{n-1} \frac{(\lambda b)^i}{i!}. \quad (2.1)$$

This formula may be used to compute an integral of the form $\int_a^b t^n e^{-\lambda t} dt$, as it is equal to $(n!/\lambda^{n+1})P[a < S_{n+1} \leq b]$. For example,

$$\int_1^2 t^3 e^{-2t} dt \;=\; \frac{3!}{2^4} P[1 < S_4 \le 2] = \frac{3}{8}[e^{-2}\sum_{i=0}^{3} 2^i/i! - e^{-4}\sum_{i=0}^{3} 4^i/i!]$$

$$=\; 0.1589.$$

In particular, when $a = 0$ and $b = \infty$, $\int_0^\infty t^n e^{-\lambda t} dt = \lambda^{-n-1} n!$.

Theorem 2.6 *(an alternative characterization of Poisson processes)*
Suppose $T_1, T_2, T_3, \ldots$ are iid $Exp(\lambda)$. Let $S_n = T_1 + \cdots + T_n$ for $n \ge 1$, and define $N(t)$ to be the number of S_n's that are $\le t$. Then $N(t)$ is a Poisson process of rate λ.

Proof: Let F_n be the distribution (function) of S_n. For $s < t$,

$$P[N(s) = i,\; N(t) - N(s) = j]$$

$$= P(S_i \le s < S_i + T_{i+1},\; S_{i+j} \le t < S_{i+j+1})$$

$$= \int_0^s P(s - u < T_{i+1},\; \sum_{k=1}^{j} T_{i+k} \le t - u < \sum_{k=1}^{j} T_{i+k} + T_{i+j+1}) dF_i(u)$$

$$= \int_0^s P(s - u < T_1,\; S_j \le t - u < S_{j+1}) dF_i(u) \quad (T_i \text{ are iid})$$

$$= \int_0^s P[T_1 - (s - u) + \sum_{k=2}^{j} T_k \le t - s < T_1 - (s - u) + \sum_{k=2}^{j+1} T_k$$

$$| T_1 > s - u]\, P(T_1 > s - u) dF_i(u)$$

$$= \int_0^s P(S_j \le t - s < S_{j+1})\, P(T_1 > s - u) dF_i(u)$$

(the lack of memory property of T_1)

$$= \int_0^s P(T_1 > s - u) dF_i(u)\, P[N(t - s) = j]$$

$$= \int_0^s P(T_{i+1} > s - u) dF_i(u)\, P[N(t - s) = j]$$

$$= P(S_i \le s < S_i + T_{i+1})\, P[N(t - s) = j]$$

$$= P[N(s) = i]\, P[N(t - s) = j].$$

Summing over i, we get $P[N(t) - N(s) = j] = P[N(t - s) = j]$. This shows that $N(t) - N(s) \stackrel{d}{=} N(t - s)$ and hence $N(t)$ has stationary increments. Then the above computation also shows that

$N(s)$ and $N(t) - N(s)$ are independent. In this computation, replacing $[N(s) = i]$ by $[N(s_1) = i_1, N(s_2) = i_2, \ldots, N(s_p) = i_p, N(s) = i]$ for $s_1 < s_2 < \cdots < s_p < s$ and nonnegative integers $i_1 \le i_2 \le \cdots \le i_p \le i$, and replacing $F_i(x) = P(S_i \le x)$ by $F(x) = P[N(s_1) = i_1, N(s_2) = i_2, \ldots, N(s_p) = i_p, S_i \le x]$, shows that $N(t) - N(s)$ is independent of $N(s_1), N(s_2), \ldots, N(s_p), N(s)$, and hence $N(t)$ has independent increments.

It remains to show that $N(t)$ has a Poisson distribution of mean λt. By the assumption, S_n is Erlang(n, λ). For any integer $n \ge 0$,

$$P[N(t) = n] = P(S_n \le t < S_{n+1}) = \int_0^t P(T_{n+1} > t - u)dF_n(u)$$

$$= \int_0^t e^{-\lambda(t-u)} \frac{\lambda^n u^{n-1}}{(n-1)!} e^{-\lambda u} du = e^{-\lambda t} \frac{\lambda^n}{(n-1)!} \int_0^t u^{n-1} du$$

$$= e^{-\lambda t} \frac{(\lambda t)^n}{n!}. \quad \diamondsuit$$

Example 2.7 Identical machine parts with an exponential lifetime of mean 0.5 are successively put in operation, and each is replaced upon failure by a new part. Find:
(a) P(at most 1 replacement by time 1 and at most 2 by time 2);
(b) P(3rd replacement occurs between times 1 and 2, and 4th within 1/2 time).

Solution: Let $N(t)$ be the number of replacements by time t. Then $N(t)$ is a Poisson process of rate $1/0.5 = 2$ with successive replacement times S_n as arrival times.
(a) $P[N(1) = 0, N(2) - N(1) \le 2] + P[N(1) = 1, N(2) - N(1) \le 1]$
$= e^{-2}e^{-2}(1 + 2 + 2^2/2) + (e^{-2}2)e^{-2}(1 + 2) = 0.2015.$
(b) $P(1 < S_3 < 2, T_4 < 0.5) = P(1 < S_3 < 2)P(T_4 < 0.5)$
$= [e^{-2}(1 + 2 + 2^2/2) - e^{-4}(1 + 4 + 4^2/2)](1 - e^{-2(0.5)}) = 0.339.$

Exercise 2.3 Customers arrive at a store according to a Poisson process at a rate of one every 5 minutes. The store opens at 8 a.m. Find
(a) P(2nd customer arrives before 8:15 and 3rd within 5 minutes);
(b) P(2nd customer arrives before 8:15 and 3rd before 8:20);
(c) P(3rd customer arrives between 8:15 and 8:20).

Exercise 2.4 A long line of customers are waiting for service at a single-server facility. Each customer will require an exponential random service time of mean 2. Suppose the service begins at time zero. Find

(a) P(first five customers complete service by time 10);
(b) P(service of 3rd customer in line begins before time 5 and is completed before time 10).

Exercise 2.5 Suppose emergency calls and non-emergency calls arrive at a call center according to two independent Poisson processes. On average, there are ten non-emergency calls and two emergency calls per hour. Find the probability that at least five non-emergency calls arrive before the first emergency call.

2.3 Conditional distribution of arrival times

Order statistics: Let $X_1, X_2, \cdots, X_n$ be n iid continuous random variables. Their ordered values $Y_1 < Y_2 < \cdots < Y_n$ are called the order statistics of $X_1, X_2, \ldots, X_n$.

If $X_1, X_2, \ldots, X_n$ are iid $U(0, t)$ for some $t > 0$, then

$$P(t_1 < Y_1 < t_1 + h, t_2 < Y_2 < t_2 + h, \ldots, t_n < Y_n < t_n + h)$$
$$= n! P(t_1 < X_1 < t_1 + h, t_2 < X_2 < t_2 + h, \ldots, t_n < X_n < t_n + h)$$
$$= \frac{n! h^n}{t^n} \tag{2.2}$$

for $0 \le t_1 < t_1 + h < t_2 < t_2 + h < \cdots < t_n < t_n + h \le t$.

Theorem 2.8 *Let $N(t)$ be a Poisson process. Fix $t > 0$. Then given $N(t) = n > 0$, the first n arrival times $S_1 < S_2 < \cdots < S_n$ are the order statistics of n iid $U(0, t)$. Consequently, for $0 < a < t$, given $N(t) = n > 0$, $N(a)$ has binomial distribution $B(n, a/t)$.*

Proof: For $0 \le t_1 < t_2 < \cdots < t_n < t_{n+1} = t$ and $h > 0$ sufficiently small,

$$P[t_i \le S_i < t_i + h, i = 1, 2, \ldots, n \mid N(t) = n]$$
$$= \frac{P\{\begin{array}{l} \text{exactly 1 arrival in } [t_i, t_i + h], 1 \le i \le n, \\ \text{and 0 elsewhere in } [0, t] \end{array}\}}{P[N(t) = n]}$$
$$= \frac{(\lambda h e^{-\lambda h})^n e^{-\lambda(t - nh)}}{e^{-\lambda t}(\lambda t)^n / n!} = \frac{n! h^n}{t^n}.$$

This shows that given $N(t) = n$, $S_1, S_2, \ldots, S_n$ have the same joint distribution as the order statistics of n iid $U(0,t)$. ◇

Example 2.9 Find $P[S_2 < 1 \mid N(4) = 5]$.

Solution: $P[S_2 < 1 \mid N(4) = 5] = P[N(1) \geq 2 \mid N(4) = 5] = 1 - [(3/4)^5 + 5(1/4)(3/4)^4] = 0.3672$ (because $N(1)$ is $B(5, 1/4)$ given $N(4) = 5$). Alternatively,

$$P(S_2 < 1 \mid N(4) = 5) = \frac{P[N(1) \geq 2, N(4) = 5]}{P[N(4) = 5]}$$

$$= \frac{P[N(4) = 5] - P[N(1) = 0, N(1,4] = 5] - P[N(1) = 1, N(1,4] = 4]}{P[N(4) = 5]}$$

$$= \frac{e^{-4\lambda}(4\lambda)^5/5! - e^{-\lambda}e^{-3\lambda}(3\lambda)^5/5! - e^{-\lambda}\lambda e^{-3\lambda}(3\lambda)^4/4!}{e^{-4\lambda}(4\lambda)^5/5!}$$

$$= \frac{4^5/5! - 3^5/5! - 3^4/4!}{4^5/5!} = 0.3672.$$

Example 2.10 (mean sum of arrival times) Let $N(t)$ be a Poisson process of rate λ with successive arrival times $S_1, S_2, \ldots$. Then the mean sum of arrival times by time t is

$$E[\sum_{n=1}^{N(t)} S_n] = E\{E[\sum_{n=1}^{N(t)} S_n \mid N(t)]\}$$

$$= E\{E[\sum_{i=1}^{n} X_i] \mid_{n=N(t)}\} \quad (X_i \text{ are iid } U(0,t))$$

$$= E\{N(t)\frac{t}{2}\} = \frac{\lambda t^2}{2}, \qquad (2.3)$$

where $E[\sum_{i=1}^{n} X_i] \mid_{n=N(t)}$ is obtained by first computing $E[\sum_{i=1}^{n} X_i]$ for an arbitrary n and then substituting $N(t)$ for n.

By the symmetry of the uniform distribution,

$$E[\sum_{n=1}^{N(t)} (t - S_n)] = E\{E[\sum_{i=1}^{n} (t - X_i)] \mid_{n=N(t)}\} = E\{E[\sum_{i=1}^{n} X_i]_{n=N(t)}\}$$

$$= \frac{\lambda t^2}{2}. \qquad (2.4)$$

Alternatively, $E[\sum_{n=1}^{N(t)} (t - S_n)] = E[\int_0^t N(s)ds] = \int_0^t E[N(s)]ds = \int_0^t \lambda s ds = \lambda t^2/2$.

Theorem 2.11 Let $t > 0$ and $n > 1$. Then given $[S_n = t]$, the first $n - 1$ arrival times $S_1 < S_2 < \cdots < S_{n-1}$ are order statistics of $n - 1$ iid random variables uniformly on $(0, t)$. Consequently, for $0 < a < t$, given $S_n = t$, $N(a)$ has the binomial distribution $B(n - 1, a/t)$.

Proof: Given $[S_n = t]$ means that $[N(t) = n]$ is given together with an arrival at time t. Because given $N(t) = n$, the first n arrivals are iid uniformly on $(0, t)$; thus, if one of them is to occur at time t, the other $n - 1$ arrivals should still be iid uniformly on $(0, t)$.

This can also be proved more formally as follows. For $0 \leq t_1 < t_2 < \cdots < t_n = t$ and $h > 0$ sufficiently small, by (1.22),

$$P[t_i \leq S_i < t_i + h \text{ for } i = 1, 2, \ldots, n - 1 \mid S_n = t]$$

$$= \lim_{\varepsilon \to 0} P[t_i \leq S_i < t_i + h \text{ for } i = 1, 2, \ldots, n - 1 \mid t - \varepsilon < S_n \leq t]$$

$$= \lim_{\varepsilon \to 0} \frac{P\{\begin{array}{l} 1 \text{ arrival in } [t_i, t_i + h], 1 \leq i \leq n - 1, \text{ and } 1 \text{ in } (t - \varepsilon, t], \\ \text{and } 0 \text{ elsewhere in } [0, t] \end{array}\}}{\sum_{i=1}^{n} P\{n - i \text{ arrivals in } [0, t - \varepsilon] \text{ and } i \text{ in } (t - \varepsilon, t]\}}$$

$$= \lim_{\varepsilon \to 0} \frac{(\lambda h e^{-\lambda h})^{n-1} e^{-\lambda \varepsilon} (\lambda \varepsilon) e^{-\lambda[t - (n-1)h - \varepsilon]}}{\sum_{i=1}^{n} [e^{-\lambda(t-\varepsilon)} (\lambda(t - \varepsilon))^{n-i}/(n - i)!][e^{-\lambda \varepsilon} (\lambda \varepsilon)^i/i!]}$$

$$= \frac{(n - 1)! h^{n-1}}{t^{n-1}} \quad \text{(the joint probability of order statistics).} \quad \Diamond$$

Example 2.12 For a Poisson process $N(t)$ of rate λ, find

$$E[N(1) = 2, 2 < S_3 < 3].$$

Solution:

$$P[N(1) = 2, 2 < S_3 < 3] = \int_2^3 P[N(1) = 2 \mid S_3 = t] dF_{S_3}(t)$$

$$= \int_2^3 [(\frac{1}{t})^2] \frac{\lambda^3 t^2}{2} e^{-\lambda t} dt = \frac{\lambda^2}{2} \int_2^3 \lambda e^{-\lambda t} dt = \frac{\lambda^2}{2} (e^{-2\lambda} - e^{-3\lambda}).$$

Alternatively,

$$P[N(1) = 2, 2 < S_3 < 3]$$
$$= P[N(1) = 2] P[N(1, 2] = 0] P[N(2, 3] > 0]$$
$$= (e^{-\lambda} \lambda^2/2)(e^{-\lambda})(1 - e^{-\lambda}) = (\lambda^2/2)(e^{-2\lambda} - e^{-3\lambda}).$$

Exercise 2.6 Find $P[S_5 > 1.5 \mid N(2) = 6]$.

Exercise 2.7 Suppose customers arrive at a system according to a Poisson process of rate 5. Each customer stays in the system for a random time uniformly between 0 and 1. Suppose each customer pays a fee equal in value to the time he spends in the system, and at time 4, all customers must leave and pay their fees. Find the expected total fee collection at time 4.

Exercise 2.8 Suppose incomes, each in the amount A, come into a company according to a Poisson process of rate λ. Suppose these incomes are immediately invested at a continuous interest rate r so that each dollar invested will grow to e^{rt} after t units of time. Find the expected total accumulation by time t.

Exercise 2.9 For a Poisson process of rate $\lambda = 2$, find

$$E[N(S_4) - N(1) \mid S_4 > 1] \quad \text{and} \quad E[N(S_4 - 1) \mid S_4 > 1].$$

Note: The above two expressions should differ by 1. This is of course no coincidence and is a consequence of the following more general result: For any integer $n > 1$ and real $a > 0$, given $[S_n > a]$, $n - 1 - N(a)$ and $N(S_n - a)$ have the same distribution.

Exercise 2.10 Establish the following extension of Theorem 2.8 (see [11, section 3.3]): Let $N(t)$ be a Poisson process of rate λ. For any interval $I = (a, b]$, let $|I| = b - a$ and $N(I) = N(a, b]$. Fix $t > 0$ and let $I_1, I_2, \ldots, I_k$ be disjoint subintervals of $[0, t]$. Then for any nonnegative integers $n_1, n_2, \ldots, n_k$ with $n = n_1 + n_2 + \cdots + n_k$,

$$P[N(I_1) = n_1, \ldots, N(I_k) = n_k \mid N(t) = n] = \frac{n!}{n_1! \cdots n_k!} p_1^{n_1} \cdots p_k^{n_k},$$

where $p_i = |I_i|/t$ for $1 \leq i \leq k$. A similar extension holds for Theorem 2.11.

2.4 Poisson processes with different types of events

Different types of events: Let $N(t)$ be a Poisson process of rate λ. Suppose its events are classified into k different types, and the classification of an event depends only on the time when it occurs and is independent of the other events. Let

$$P_i(t) = P[\text{event is type } i \mid \text{it occurs at time } t] \qquad (2.5)$$

and let $N_i(t)$ be the numbers of type i events by time t for $i = 1, 2, \ldots, k$.

Theorem 2.13 *For each $t > 0$, $N_i(t)$ are independent Poisson random variables of means $\lambda t p_i$, where*

$$p_i = (1/t) \int_0^t P_i(s)ds \qquad \text{(time average of type } i \text{ probability).} \qquad (2.6)$$

Proof: For simplicity, we will only prove for $k = 2$. Fix two nonnegative integers n and m, Given $N(t) = n + m$, the arrival times $S_1 < S_2 < \cdots < S_{n+m}$ by time t are order statistics of iid $U_1, U_2, \ldots, U_{n+m}$ with uniform distribution $U(0, t)$. For $1 \le j \le n+m$, let $X_j = 1$ if the event at time U_j is type 1, and let $X_j = 0$ if otherwise. Then

$$P(X_j = 1) = E[P(X_j = 1 \mid U_j)] = E[P_1(U_j)] = \int_0^t P_1(s)ds/t = p_1.$$

Write p for p_1. Then $p_2 = 1 - p$ and

$$\begin{aligned}
&P[N_1(t) = n, N_2(t) = m] \\
&= P[N_1(t) = n, N_2(t) = m \mid N(t) = n + m]P[N(t) = n + m] \\
&= \binom{n+m}{n} p^n (1 - p)^m e^{-\lambda t}\frac{(\lambda t)^{n+m}}{(n + m)!} \\
&= e^{-\lambda pt}\frac{(\lambda tp)^n}{n!} \cdot e^{-\lambda(1-p)t}\frac{(\lambda(1 - p)t)^m}{m!}. \qquad (2.7)
\end{aligned}$$

Summing over m in (2.7) yields $P[N_1(t) = n] = e^{-\lambda pt}(\lambda pt)^n/n!$ and proves that $N_1(t)$ is Poisson of mean λpt. Similarly summing over n in (2.7) will prove that $N_2(t)$ is Poisson of mean $\lambda(1 - p)t$. Now (2.7) implies the independence of $N_1(t)$ and $N_2(t)$. $\diamond$

Note: If $P_i(t)$ is not a constant, then $N_i(t)$ is not a Poisson process.

Example 2.14 Starting at time 0, visitors enter a museum according to a Poisson process of rate $\lambda = 2$. Each visitor spends a random time in the museum that is uniformly between 0 and 1. Find:
(a) the expected number of visitors in the museum at time 5;
(b) the probability that six have entered and two still remain in the museum at time 5.

Solution: (a) A visitor is classified as type 1 if he is in the museum at time 5. Then

$$P_1(t) = P[\text{visitor arriving at time } t \text{ still in museum at time 5}]$$
$$= \begin{cases} 0, & \text{if } t < 4 \\ t - 4, & \text{if } 4 < t < 5. \end{cases}$$

Then $p_1 = (1/5) \int_0^5 P(t)dt = (1/5) \int_4^5 (t-4)dt = (1/5) \int_0^1 t\, dt = 1/10$ and $E[N_1(5)] = (2)(5)(1/10) = 1$.

(b) $N_1(5)$ and $N_2(5)$ are independent Poisson of means $2(5)(1/10) = 1$ and $2(5)(9/10) = 9$. Then $P[N_1(5) = 2, N(5) = 6]$
$= P[N_1(5) = 2]P[N_2(5) = 4] = (e^{-1}/2)(e^{-9} \cdot 9^4/4!) = 0.0062$.

Exercise 2.11 Starting at time 0, customers enter a service facility according to a Poisson process of rate 5. Each customer stays an exponential random time of mean 2 in the facility and then leaves.
(a) Find the expected number of customers in the facility at time 2.
(b) Find $P(\text{five in facility and four have left at time 2})$.

Exercise 2.12 Passengers enter a train station according to a Poisson process of rate 4. A train departs the station at a random time uniformly distributed between 0 and 5, and a second train departs at time 5.
(a) Find the expected number of passengers who wait for 1 unit of time or less.
(b) Find $P(\text{at least three who wait more than 1 unit of time})$.

2.5 Compound Poisson processes

Definition and basic formulas: Let $N(t)$ be a Poisson process of rate λ and let $Y_1, Y_2, \ldots,$ be iid random variables which are independent of the Poisson process. The process

$$X(t) = \sum_{i=1}^{N(t)} Y_i \tag{2.8}$$

is called a compound Poisson process.

Theorem 2.15 *For a compound Poisson process $X(t)$ defined above,*

$$E[X(t)] = \lambda t E(Y), \quad \mathrm{Var}[X(t)] = \lambda t E(Y^2), \quad L_{X(t)}(s) = e^{-\lambda t + \lambda t L_Y(s)}.$$

Proof: The first equality above follows directly from Wald's identity. The proofs of the second and third equalities are left as Exercise 2.14.
◇

Example 2.16 Insurance claims occur according to a Poisson process of rate $\lambda = 10$ per month and the amounts of claims are iid of mean 500 and standard deviation 200. Find the expected total claims in a three-month period. Based on the normal approximation, estimate the probability that the total claim will not exceed 20,000.

Solution: Let $N(t)$ be the Poisson process that counts the number of claims and let Y_i be the iid claim amounts. Then the total amount $X(t)$ of claims by time t is a compound Poisson process given in (2.8). The expected total claim is $E[X(3)] = (10)(3)(500) = 15,000$. The variance is $(10)(3)(200^2 + 500^2) = 8,700,000$. Using the normal distribution N(15,000, 8,700,000) with Z being a standard normal random variable, the probability of not exceeding 20,000 is

$$P[X(3) \le 20,000] \approx P(Z \le \frac{20,000 - 15,000}{\sqrt{8,700,000}})$$
$$= P(Z \le \frac{5,000}{2950}) = P(Z \le 1.7) = 0.9554$$

from a standard normal distribution table.

Exercise 2.13 Buses arrive at a park according to a Poisson process of rate 5 per hour beginning at 8 a.m. Each bus carries between 1 to 10 visitors of equal probability. Find the expected number and the standard deviation of visitors who arrive at the park by 10 a.m.

Exercise 2.14 Let $N(t)$ be a Poisson process of rate λ, let Y_i are iid with Laplace transform $L_Y(s)$ and let $X(t)$ be a compound Poisson process defined by (2.8).
(a) Find the Laplace transform $L_{X(t)}(s)$ of $X(t)$.
(b) Use $L_{X(t)}(s)$ to find $\mathrm{Var}[X(t)]$.

2.6   Nonhomogeneous Poisson processes

Definition: A counting process $N(t)$ with $N(0) = 0$ is called a nonho-mogeneous Poisson process if it has independent (but not necessarily stationary) increments, and for $s < t$, $N(t) - N(s)$ has a Poisson distribution of mean $m(t) - m(s)$, where $m(t)$ is a continuous increasing function on $\mathbb{R}_+ = [0, \infty)$ with $m(0) = 0$. Here a Poisson distribution of mean 0 is defined as the distribution concentrated at 0.

The function $m(t)$ is called the mean function of the nonhomogeneous Poisson process $N(t)$ because $m(t) = E[N(t)]$.

In many applications, the mean function $m(t)$ possesses a derivative $\lambda(t)$ such that

$$m(t) = \int_0^t \lambda(s)ds. \tag{2.9}$$

In this case, $\lambda(t)$ is called the rate function which must be ≥ 0 because $m(t)$ is increasing. A Poisson process of rate λ is just a nonhomogeneous Poisson process with a constant rate function $\lambda(t) = \lambda$, which may also be called a homogeneous Poisson process.

Fixed jumps: A counting process $N(t)$ is said to have a fixed jump at time $t > 0$ if

$$P[N(t) - N(t-) > 0] > 0,$$

that is, if there is a positive probability that an event occurs at time t. Otherwise, if $P[N(t) - N(t-)] = 0$ for all $t > 0$, $N(t)$ will be said to have no fixed jump.

We note that a nonhomogeneous Poisson process $N(t)$ has no fixed jump because by the continuity of $m(t)$,

$$
\begin{aligned}
P[N(t) - N(t-) > 0] &= \lim_{s \uparrow t} P[N(t) - N(s) > 0] \\
&= \lim_{s \uparrow t} [1 - e^{-[m(t)-m(s)]}] = 0.
\end{aligned}
$$

Application of nonhomogeneous Poisson processes: We will see in Theorem 2.17 that any simple counting process with independent increments and no fixed jumps is a nonhomogeneous Poisson process. Therefore, the number of events occurring successively along the time line, with rate of occurrence changing in time, may be modeled by a nonhomogeneous Poisson process if the numbers of events in non-overlapping intervals are independent.

Conversion to a Poisson process: Let $N(t)$ be a nonhomogeneous Poisson process with mean function $m(t)$, and let $m^{-1}(t)$ be the inverse function of $m(t)$. Then $\tilde{N}(t) = N(m^{-1}(t))$, a time-changed process, is a homogeneous Poisson process of rate 1.

This can be easily proved by checking that for $t > s$, $\tilde{N}(t) - \tilde{N}(s)$ has a Poisson distribution of mean $t - s$. Therefore, a nonhomogeneous Poisson process is a homogeneous Poisson process up to a time change.

When $m(t)$ is not strictly increasing, its inverse $m^{-1}(t)$ may not exist in the usual sense, but may be defined by

$$m^{-1}(u) = \inf\{t > 0; \ m(t) > u\}. \tag{2.10}$$

Then $m(m^{-1}(t)) = t$ for all $t \geq 0$, and hence $\tilde{N}(t) = N(m^{-1}(t))$ is still a Poisson process of rate 1.

Because a Poisson process is a simple counting process, it follows that a nonhomogeneous Poisson process is also simple.

Theorem 2.17 *Let $N(t)$ be a simple counting process with $N(0) = 0$. Assume $N(t)$ has independent increments and no fixed jumps, and $m(t) = E[N(t)]$ is finite. Then $N(t)$ is a nonhomogeneous Poisson process with mean function $m(t)$.*

Proof: For $s < t$, divide the time interval $(s, t]$ into n subintervals of equal length, in the form of $(s + (j-1)(t-s)/n, \ s + j(t-s)/n]$ for $j = 1, 2, \ldots, n$. Let p_{nj} be the probability that an event occurs in the jth subinterval. Note that

$$
\begin{aligned}
p_{nj} &\leq E[N(s + \frac{j(t-s)}{n}) - N(s + \frac{(j-1)(t-s)}{n})] \\
&= m(s + \frac{j(t-s)}{n}) - m(s + \frac{(j-1)(t-s)}{n}).
\end{aligned}
$$

No fixed jump means that $m(t) = E[N(t)]$ is continuous. Then $\max_{1 \leq j \leq n} p_{nj} \to 0$ as $n \to \infty$. Because as a simple counting process, no two events occur at same time, and hence

$$\sum_{j=1}^{n} p_{nj} = E[\text{\# of subintervals containing an event}]$$

$$\to \quad E[\text{\# of events in } (s, t]] = E[N(t) - N(s)] = m(t) - m(s).$$

By the law of rare events (§1.7), $N(t) - N(s)$ has a Poisson distribution of mean $m(t) - m(s)$. ◇

Arrival times: The distribution of nth arrival time S_n, $n \geq 1$, is given by

$$P(S_n > t) = P[N(t) < n]$$
$$= e^{-m(t)}[1 + m(t) + \frac{m(t)^2}{2!} + \cdots + \frac{m(t)^{n-1}}{(n-1)!}] \quad (2.11)$$

for $t \geq 0$. Let $m(\infty) = \lim_{t\to\infty} m(t)$. Note that if $m(\infty) < \infty$, then S_n is deficient, that is, $P(S_n = \infty) > 0$. In fact,

$$P(S_n = \infty) = \lim_{t\to\infty} P(S_n > t)$$
$$= e^{-m(\infty)}[1 + m(\infty) + \frac{m(\infty)^2}{2!} + \cdots + \frac{m(\infty)^{n-1}}{(n-1)!}]. \quad (2.12)$$

When there is a rate function $\lambda(t)$, the distribution of S_n has a density

$$f_n(t) = -\frac{d}{dt}P(S_n > t) = \frac{m(t)^{n-1}}{(n-1)!}e^{-m(t)}\lambda(t), \quad 0 < t < \infty, \quad (2.13)$$

which may be used in probability computation, but one needs to remember that when $m(\infty) < \infty$, $\int_0^\infty f_n(t)dt = P(S_n < \infty) < 1$.

Theorem 2.18 (*Conditional distribution of arrival times*) *Given* $N(t) = n$, *the first n arrival times $S_1, \ldots, S_n$ are identical in distribution to the order statistics of n iid random variables with the following distribution function*

$$F(x) = \begin{cases} 0, & x < 0; \\ m(x)/m(t), & 0 \leq x \leq t; \\ 1, & x > t. \end{cases} \quad (2.14)$$

Consequently, for $s < t$, given $N(t) = n$, $N(s)$ is $B(n, m(s)/m(t))$.

Proof: Given $[N(t) = n] = [\tilde{N}(m(t)) = n]$, an un-ordered arrival time $\tilde{S}$ of the first n arrivals of the Poisson process $\tilde{N}$ has uniform distribution on $[0, m(t)]$, that is, $P[\tilde{S} \leq x] = x/m(t)$ for $0 \leq x \leq m(t)$, where P denotes the conditional probability given $N(t) = n$. Let S be an un-ordered arrival time of the first n arrivals of the nonhomogeneous process $N(t)$. First assume $m(t)$ is strictly increasing so that its inverse $m^{-1}(t)$ exists in the usual sense. Then $S = m^{-1}(\tilde{S})$, and for $0 \le y \le t$,

$$P[S \le y] = P[m^{-1}(\tilde{S}) \le y] = P[\tilde{S} \le m(y)] = \frac{m(y)}{m(t)}.$$

This proves (2.14). If $m(t)$ is not strictly increasing, then the generalized inverse $m^{-1}(t)$ defined by (2.10) is not continuous and it jumps to skip any interval where $m(t)$ is a constant, and during such an interval, no arrival can occur. Then S is not a value skipped by $m^{-1}(t)$ and hence $S = m^{-1}(\tilde{S})$ holds. The above computation is still valid. $\diamond$.

Example 2.19 Let $N(t)$ be a nonhomogeneous Poisson process with mean function $m(t) = 4t + t^2$ and let S_n be its successive arrival times.
(a) Find $P(S_2 > 1)$.
(b) Find $P[S_2 > 1 \mid N(2) = 4]$.

Solution: (a) $P(S_2 > 1) = P[N(1) \le 1] = e^{-m(1)}[1 + m(1)] = e^{-5}(1 + 5) = 0.0404$.
(b) $P[S_2 > 1 \mid N(2) = 4] = P[N(1) \le 1 \mid N(2) = 4] = (7/12)^4 + 4(5/12)(7/12)^3 = 0.4466$ because given $N(2) = 4$, $N(1)$ is binomial $B(n, p)$ with $n = 4$ and $p = m(1)/m(2) = 5/12$.

Exercise 2.15 For a nonhomogeneous Poisson process with rate function $\lambda(t) = 1 + (1/5)t$, find

$$P(S_2 < 2 < S_4) \quad \text{and} \quad P[S_2 < 2 < S_4 \mid N(4) = 5].$$

Exercise 2.16 Visitors enter a museum according to a nonhomogeneous Poisson process with rate function given in Exercise 2.15. Each visitor spends a random amount time in the museum, which is uniformly distributed between 0 and 1. Find the expected number of visitors in the museum at time 5.
Hint: Convert to a homogeneous Poisson process of unit rate by a time change.

Exercise 2.17 The inter-arrival times for a nonhomogeneous Poisson process are defined just as for a Poisson process; let $T_1 = S_1$ and $T_{n+1} = S_{n+1} - S_n$ for $n \ge 1$, but set $T_n = \infty$ if $S_n = \infty$. Show that if T_n are independent, then $N(t)$ is a homogeneous Poisson process.
Hint: Look at $P(T_{n+1} > t \mid S_n = x)$.

Chapter 3

Renewal processes

3.1 An introduction to renewal processes

Renewal processes: Let T_n, for $n \geq 1$, be a sequence of iid nonnegative random variables with the common distribution (function) $F(t)$ satisfying $F(0) = P(T_n = 0) < 1$. Let

$$S_n = T_1 + \cdots + T_n,$$

called the nth renewal time, and set $S_0 = 0$. The sequence S_n is called a renewal sequence. The time interval $(S_{n-1}, S_n]$ between two successive renewals is called a renewal cycle and its length is T_n.

Let

$$N(t) = \#\{n \geq 1; \ S_n \leq t\}$$

be the number of the renewals by time t. This is a counting process as defined in §2.1 and is called a renewal counting process or simply a renewal process.

By Theorems 2.5 and 2.6, a Poisson process is a renewal process with an exponential cycle time distribution F, and conversely, a renewal process with F being $\text{Exp}(\lambda)$ is a Poisson process of rate λ.

Note that $P(T_n = 0) > 0$ is allowed, so that it is possible to have several renewals occurring at the same time, and $N(0)$ may not be 0. However, there can only be a finite number of renewals at any time because this number plus 1 has a geometric distribution with $p = P(T_n > 0)$ being the probability of success.

Basic relations between S_n and $N(t)$: For $n \geq 0$,

$$\begin{aligned}
[N(t) \leq n] &= [S_{n+1} > t], \\
[N(t) = n] &= [S_n \leq t < S_{n+1}], \\
S_{N(t)} &\leq t < S_{N(t)+1}.
\end{aligned} \tag{3.1}$$

Delayed renewal processes: Sometimes it may be useful to slightly extend the definition of renewal processes by allowing the initial renewal time T_1 to have a distribution $G(t)$, that is different from $F(t)$, but still independent of the rest of T_n's. Then the associated counting process is called a delayed renewal process and is denoted by $N_D(t)$. In contrast, the renewal process defined earlier may be called a pure renewal process.

For example, if the renewals of a pure renewal process are counted after the first renewal cycle has already started, we get a delayed renewal process, but not all delayed renewal processes are obtained this way. Most of our results will be stated and proved for pure renewal processes, but many of them hold also for delayed renewal processes.

Example 3.1 Suppose an item such as a light bulb is installed at time 0. At the end of its lifetime, it is replaced by a new identical item (meaning its lifetime has the same probability distribution), and this process continues indefinitely. Then the lifetimes T_n of the items are iid, the successive replacement times S_n form a renewal sequence, and the associated renewal process $N(t)$ is the number of replacements by time t. If at time 0, an item is already installed for some time, then one obtains a delayed renewal process.

Mean function: Let $m(t) = E[N(t)]$. This is the mean number of renewals by time t and is called the mean function or the renewal function.

Let $\mu = E(T_n)$, the mean cycle time. In the delayed case, one assumes $n > 1$ in this definition. In the sequel, T will denote a generic T_n and in the delayed case, a T_n with $n > 1$, unless explicitly stated otherwise.

The following relation can be easily established. For a pure renewal process,

$$m(t) = \sum_{n=1}^{\infty} P(S_n \leq t) = \sum_{n=1}^{\infty} F_n(t), \qquad (3.2)$$

where $F_n = F^{*n}$ is the n-fold convolution of F and is the distribution function of S_n.

For a delayed renewal process, its mean function is given by

$$m_D(t) = G(t) + \sum_{n=1}^{\infty} G * F^{*n}(t) = G(t) + G * m(t). \qquad (3.3)$$

Theorem 3.2 *For a finite $t \geq 0$, $m(t) < \infty$. Consequently, $N(t) < \infty$ a.s. for finite $t \geq 0$ and $S_n \to \infty$ a.s. as $n \to \infty$. Moreover, $N(t) \to \infty$ a.s. as $t \to \infty$. The results also hold for a delayed renewal process.*

Proof: Because $F(0) < 1$, there is $b > 0$ with $p = P(T \geq b) > 0$. Choose an integer j with $t \leq jb$. Consider a sequence of independent trials based on T_n with two outcomes at each trial: success if $T_n \geq b$ and failure if otherwise. Let Z be the number of trials until the jth success. Then $E(Z) < \infty$ (considering the sum of j iid geometric random variables). Since $S_Z \geq jb$, $m(t) \leq m(jb) = E[N(jb)] \leq E[N(S_Z)] = E(Z) < \infty$. For a delayed renewal process, by (3.3), $m_D(t) < \infty$.

Because $S_n < \infty$ for any finite $n > 0$, $N(t)$ must converge to ∞ as $t \to \infty$. ◇

Theorem 3.3
$$\frac{N(t)}{t} \to \frac{1}{\mu} \quad \text{a.s. as } t \to \infty. \tag{3.4}$$

This is true even when $\mu = \infty$ with $1/\infty = 0$, and true also for a delayed renewal process.

Proof: By the SLLN, $S_{N(t)}/N(t) = \sum_{n=1}^{N(t)} T_n/N(t) \to \mu$ and
$$\frac{S_{N(t)+1}}{N(t)} = \frac{S_{N(t)+1}}{N(t)+1} \cdot \frac{N(t)+1}{N(t)} \to \mu \quad \text{a.s.}$$

By (3.1), $N(t)/S_{N(t)+1} \leq N(t)/t \leq N(t)/S_{N(t)}$, it follows that $N(t)/t \to 1/\mu$. ◇

The following result says that the limit in (3.4) in fact holds in expectation.

Theorem 3.4 *(Elementary renewal theorem)*
$$\lim_{t \to \infty} \frac{m(t)}{t} = \lim_{t \to \infty} \frac{E[N(t)]}{t} = \frac{1}{\mu} \quad (\mu \leq \infty).$$

Proof: We first briefly review the limit infimum and the limit supremum of a sequence of $a_n \geq 0$, which are defined as $\liminf_{n \to \infty} a_n = \lim_{k \to \infty} \inf\{a_n; n \geq k\}$ and $\limsup_{n \to \infty} a_n = \lim_{k \to \infty} \sup\{a_n; n \geq k\}$, where the infimum $\inf\{\cdots\}$ is defined before, and $\sup\{\cdots\}$ is the supremum or the least upper bound of the nonnegative numbers in $\{\cdots\}$ (defined to be 0 when the set $\{\cdots\}$ is empty). We always have

$\liminf a_n \leq \limsup a_n$, and if $\liminf a_n = \limsup a_n$, then $\lim_{n\to\infty} a_n$ exists and is equal to this common value.

By Wald's identity, noting that $N(t)+1$ is a stopping time of the iid sequence T_n, one may take expectation of $S_{N(t)+1} = \sum_{n=1}^{N(t)+1} T_n \geq t$ to get $\mu[m(t) + 1] \geq t$, hence, $\liminf_{t\to\infty} m(t)/t \geq 1/\mu$. Next, fix a constant $c > 0$ and define a truncated renewal process $\bar{N}(t)$ with cycle times $\bar{T}_n$ by setting $\bar{T}_n = T_n \wedge c$ (recall $a \wedge b = \min(a,b)$). Let $\bar{S}_n$ be the successive renewal times associated with $\bar{T}_n$, and let $\bar{\mu} = E(\bar{T})$ and $\bar{m}(t) = E[\bar{N}(t)]$. We have $\bar{S}_{\bar{N}(t)+1} \leq t + c$, taking expectation, $\bar{\mu}[\bar{m}(t) + 1] \leq t + c$, and $\limsup_{t\to\infty}(1/t)\bar{m}(t) \leq 1/\bar{\mu}$. Since $\bar{S}_n \leq S_n$, $\bar{N}(t) \geq N(t)$ and $\bar{m}(t) \geq m(t)$, hence, $\limsup_{t\to\infty}(1/t)m(t) \leq 1/\bar{\mu} \to 1/\mu$ as $c \to \infty$. $\Diamond$

Exercise 3.1 Suppose independent trials are performed at integer times $n \geq 1$ and each trial has a probability p being a success ($0 < p < 1$). Let $N(t)$ be the number of successes by time t. Show that $N(t)$ is a renewal process, and find its cycle time distribution $F(t)$, the mean function $m(t)$, and the distribution of the nth renewal time S_n.

Exercise 3.2 Customers arrive at a single server bank according to a Poisson process of rate λ, but they will not enter the bank if the server is busy. The service times are iid of mean μ.
(a) In the long run, how many customers are served per unit time in average?
(b) In the long run, what fraction of arriving customers enter the bank?

Exercise 3.3 Shocks occur to a system according to a renewal process with iid cycle times T_n. Each shock will bring down the system for a random time and a shock that occurs in a down period will initiate a new down period (so the effects of earlier shocks disappear). Assume a down period starts at time 0 and the downtime Y is independent of arrivals of shocks. Let R be the time when the system recovers. Show that $E(R) = E(T \wedge Y)/P(T > Y)$.

Exercise 3.4 Let $N(t)$ be a renewal process with a continuous cycle time distribution. Show that if $N(t)$ has independent increments, then it must be a Poisson process.

3.2 Renewal reward processes

Definition: Given a renewal process generated by iid cycle times T_n of mean μ, let R_n be an iid sequence of random variables, called rewards earned in successive renewal cycles. The total reward by time t,

$$R(t) = \sum_{n=1}^{N(t)} R_n, \qquad (3.5)$$

is called a renewal reward process. Note that by this definition, the rewards are realized only at renewal times $S_n = T_1 + \cdots + T_n$, $n \geq 1$.

Theorem 3.5 *(Limiting properties of renewal reward processes) Assume $E(R)$ is finite. Then*

$$\lim_{t\to\infty} \frac{R(t)}{t} = \frac{E(R)}{E(T)} = \frac{E(R)}{\mu} \quad a.s. \qquad (3.6)$$

This means the long-run average reward per unit time is the ratio of the mean reward per renewal cycle to the mean cycle length. This holds even when $\mu = \infty$. Moreover, if $\mu < \infty$ and $R_n \geq 0$, then (3.6) holds even when $E(R) = \infty$.

These results hold also for a delayed $N(t)$ when R_n are iid for $n \geq 2$.

Proof: $R(t)/t = [\sum_{n=1}^{N(t)} R_n/N(t)][N(t)/t]$. By the SLLN,

$$\sum_{n=1}^{N(t)} R_n/N(t) \to E(R) \quad \text{a.s.} \ \text{as} \ t \to \infty.$$

By Theorem 3.3, $N(t)/t \to 1/\mu$ a.s. This proves $R(t)/t \to E(R)/E(T)$ a.s. as $t \to \infty$ for $E(R) < \infty$. The proof is valid even for $\mu = \infty$, and also for $E(R) = \infty$ if $\mu < \infty$ and $R_n \geq 0$. $\diamond$

Cumulative renewal reward processes: We may assume the rewards are realized not just at renewal times, but may be realized several times in a renewal cycle or even continuously. Let $r_n(t)$, $0 \leq t \leq T_n$, be the amount of net reward accumulated in the nth cycle from time S_{n-1} to time $S_{n-1}+t$ and define $r_n(t) = R_n$ for $t > T_n$, where $R_n = r_n(T_n)$ is the total net reward accumulated in the nth cycle. We will assume the sequence of processes $r_n(t)$ are iid. The total reward $R(t)$ accumulated at time t will be called a cumulative renewal reward process. Let

$$\overline{R}_n = \sup\{|r_n(t)|;\ 0 \le t \le T_n\}.$$

Theorem 3.6 *The limiting properties of the renewal reward process in Theorem 3.5, (3.6), still hold for a cumulative renewal reward process $R(t)$ provided $E(\overline{R}) < \infty$. Moreover, if $\mu < \infty$ and $r_n(t)$ is a nonnegative increasing function, then $E(\overline{R}) = \infty$ is allowed (note that in this case, $\overline{R}_n = R_n$).*

Proof: Since $R(t) = \sum_{n=1}^{N(t)} R_n + s_t$, where $|s_t| \le \overline{R}_{N(t)+1}$, it suffices to show that if $E(\overline{R}) < \infty$, then $\overline{R}_{N(t)+1}/t \to 0$ as $t \to \infty$. By SLLN, $\overline{R}_n/n \to 0$ as $n \to \infty$, thus,

$$\frac{\overline{R}_{N(t)+1}}{t} = \frac{\overline{R}_{N(t)+1}}{N(t)+1} \cdot \frac{N(t)+1}{t} \to 0 \cdot \frac{1}{\mu} = 0 \quad \text{as} \quad t \to \infty \ \Diamond.$$

Example 3.7 (An on-off system) A machine is on for a random time X_n and then is off for a random time Y_n. Assume the pairs (X_n, Y_n) are iid, but allow possible dependence between X_n and Y_n. Then an on-interval followed by an off-interval form a renewal cycle of mean length $E(X + Y)$. Consider one unit of reward is earned for each unit of on time. We obtain the long-run fraction of on time to be

$$\lim_{t \to \infty} \frac{\text{on time by time } t}{t} = \frac{E(X)}{E(X) + E(Y)}.$$

Example 3.8 (An age replacement policy) An item with lifetime distribution F is placed in service upon failure or reaching a certain age T, whichever occurs first, and then is replaced by a new one with identical lifetime distribution. The lifetimes of successive items $X_1, X_2, \ldots$ are iid with distribution F. Suppose $a > 0$ is the cost for replacing a good item and $b > 0$ is the extra cost for replacing a damaged one. To find the long-run cost per unit time, let $Y_n = X_n \wedge T$ and consider the renewal process $N(t)$ with cycle times Y_n. Let R_n be the cost in the nth renewal cycle and let $R(t) = \sum_{n=1}^{N(t)} R_n$ be the total cost by time t. Then $E(R_n) = aP(X_n > T) + (a+b)P(X_n \le T) = a + bF(T)$ and $E(Y_n) = \int_0^\infty P(Y_n > t)dt = \int_0^T P(X_n > t)dt = \int_0^T \bar{F}(t)dt$. Therefore, the long-run average cost per unit time is given by

$$c(T) = \lim_{t \to \infty} \frac{R(t)}{t} = \frac{E(R_n)}{E(Y_n)} = [a + bF(T)]/\left[\int_0^T \bar{F}(t)dt\right].$$

Given F, $c(T)$ may be minimized by differentiating with respect to T. Note that if F is exponential, then by the lack of memory property, $c(T)$ should be minimized at $T = \infty$.

Example 3.9 A system has a key component that functions for a random amount time of mean μ. When it fails, there is a probability p that the system may automatically replace it with a new one without interruption. If the replacement is not successful, then the system shuts down and it takes a random amount time of mean ν to have the component manually replaced to bring the system back to work. Assume both μ and ν are finite and positive, and $0 < p < 1$. Find the long-run fraction of time when the system is functioning.

Solution: A renewal cycle starts when a component is manually installed. Let X_n be the lifetimes of the components being successively installed (of mean μ), and let σ be the number of replacements in a cycle. Then σ is geometric with $P(\sigma = 1) = 1 - p$, and it is independent of X_n. By Wald's identity, the mean length of a working period is $E[\sum_{n=1}^{\sigma} X_n] = E(X)E(\sigma) = \mu/(1 - p)$. Since the manual replacement time has mean ν, the mean cycle length is $\mu/(1-p)+\nu$. Assuming 1 unit reward is generated for each unit of working time, then the long-run fraction of time when the system is functioning is

$$\frac{\mu/(1-p)}{\mu/(1-p)+\nu} = \frac{\mu}{\mu+\nu(1-p)}.$$

Alternatively, a renewal cycle starts when a component is installed, automatically or manually. The mean cycle length is $\mu p+(\mu+\nu)(1-p) = \mu+\nu(1-p)$ and the mean functioning time in a cycle is μ. The long-run fraction of functioning time is still $\mu/[\mu + \nu(1 - p)]$.

Example 3.10 Passengers arrive at a train station according to a renewal process with mean cycle time μ. A train leaves the station whenever there are n passengers waiting. The station incurs a cost $a > 0$ for each waiting customer per unit time and an additional cost $b > 0$ when a train is dispatched. Find the value of n that minimizes the long-run cost per unit time.

Solution: A renewal cycle starts whenever a train departs. The mean cycle length is

$$E(X_1 + X_2 + \cdots + X_n) = n\mu,$$

where X_n are successive inter-arrival times of customers. Consider the cost as reward, then the long-run cost per unit time is

$$c(n) = \frac{E[aX_2 + 2aX_3 + \cdots + (n-1)aX_n] + b}{n\mu}$$

$$= \frac{a\mu n(n-1)/2 + b}{n\mu} = \frac{a(n-1)}{2} + \frac{b}{n\mu}.$$

Solve $c'(n) = a/2 - b/(n^2\mu) = 0$ to get $n = \sqrt{2b/(a\mu)}$. This value of n minimizes $c(n)$ because $c(n) \to \infty$ as $n \to \infty$ and also as $n \to 0$.

Exercise 3.5 As in Example 3.10, but now assume the passengers arrive according to a Poisson process of rate $\lambda = 1$, and when there are n arrivals, a train is dispatched, but it takes a random time Y of mean $\nu = 1$ and variance $\sigma^2 = 1$ to depart. Find n to minimize the long-run cost per unit time.

Exercise 3.6 In Example 3.8, assume F is uniform on $[0, L]$ for some $L > 0$. Find T that minimizes the long-run cost per unit time.

Exercise 3.7 A machine breaks down periodically, and it is put to repair when its down state is detected during a random inspection. After the repair, the machine functions as before. Let X_n be the length of the nth on-period, Y_n be the time for nth repair, and Z_n be the time between $(n-1)$st and nth inspections. Assume that the three iid sequences, X_n, Y_n, and Z_n, are independent, $E(X) = 0.4$, $E(Y) = 0.2$, and Z is exponential with $E(Z) = 0.1$.
(a) Find the long-run fraction of time when the machine is on.
(b) Suppose the inspection stops during repair. Find the long-run average number of inspections per unit time.

Exercise 3.8 Demands, each for one unit in inventory, arrive according to a Poisson process of rate λ. When the inventory level hits 0, it is refilled to S units but it takes a random time of mean μ to complete the refill. Each unit of demand that is met generates a profit of P, but during the refill time, demands are lost at a cost of c per unit. The inventory holding cost is h per unit inventory per unit time. Find the net profit per unit time in the long run.

Exercise 3.9 A system has two parallel components. One has an exponential lifetime distribution of mean 1 and the other uniform on $[0, 2]$, independently of each other. The two components are installed at the same time and will be replaced together when both are broken. Find the long-run fraction of time when only one component is working, and that when only the exponential component is working.

Exercise 3.10 Passengers arrive at a bus station according to a Poisson process at a rate of 4 every 10 minutes. A bus is dispatched every 10 minutes or whenever there are 4 passengers at the station, whichever occurs first.

(a) Find the long-run fraction of time when there are 3 passengers waiting in the station.

(b) In the long run, what fraction of passengers wait more than 5 minutes?

3.3 Queuing systems

A queuing system: A queuing system is a service station at which customers arrive according to a counting process. An arriving customer is served by one of the available servers and will then depart the system. If all servers are busy, the arriving customer will wait in queue and will be served according to some rule, such as FCFS (first come and first served). The service times are random. The servers are called homogeneous if their service times have the same distribution, otherwise, the service times of different servers may have different distributions and then they will be called heterogeneous. In the case of heterogeneous servers, a rule is in force to decide which server is to be engaged when two or more servers are available to serve an incoming customer.

In the rest of this chapter, we will assume that customers arrive at a queuing system according to a renewal process and served by the FCFS rule, and the service times are independent of each other and of the arrivals, unless when explicitly stated otherwise. We will also assume that both the mean inter-arrival time of customers and the mean service time are finite and positive, and an incoming customer is served immediately when there is an available server.

Such a system with k homogeneous servers is denoted as $G/G/k$, where the first G means a general inter-arrival time distribution, and the second G means a general service time distribution.

If the inter-arrival or service time distribution is exponential, then the letter G is to be replaced by M, which stands for the Markov property, a consequence of the exponential distribution, to be discussed later in Chapter 5. For example, $M/G/k$ is a queuing system with an exponential inter-arrival distribution and a general service time distribution with k servers. In this system, the customers arrive according to a Poisson process.

We will develop here some general properties of queuing systems. More explicit results are obtained for $M/M/k$ systems in Chapter 5, where heterogeneous exponential servers, and queuing systems with nonhomogeneous Poisson arrivals and services, are also discussed.

Stability: Assume at time 0, a customer arrives when all servers are free. Let S_n for $n = 1, 2, 3, \ldots$ be the successive times when an arriving customer finds all servers free, and let $T_n = S_n - S_{n-1}$, which are iid. Write T for a generic T_n with $n \geq 1$. The queuing system is called stable if $E(T) < \infty$. Only a stable queuing system may be expected to periodically (in a probabilistic sense) clear all customers in system. In a stable queuing system, T_n form the cycle times of a renewal process, called the renewal process of the stable queuing system, which should not be confused with the renewal process of the arrivals. A renewal cycle consists of a busy period, during which at least one server is busy, followed by an idle period, during which all servers are free. The system probabilistically repeats itself at the beginning of each renewal cycle.

The stability of a queuing system may be defined in a more general sense by allowing S_n to be any renewal sequence such that the queuing system observed after time S_n has the same distribution for all n, and is independent of $\{S_1, S_2, \ldots, S_n\}$ and the system up to time S_n. Then a stable queuing system does not have to periodically clear all customers. However, in the rest of this book, the stability is understood in the sense defined in the last paragraph.

Number of services in a cycle: Let σ be the number of services completed during a renewal cycle of a stable queuing system and let X_n be the successive inter-arrival times of customers in the cycle. Then the cycle length is $T = \sum_{n=1}^{\sigma} X_n$. Because for any integer $n > 0$, the event $[\sigma = n]$ depends only on $X_1, X_2, \ldots, X_n$ and the first n services, σ is a stopping time of X_n. By Wald's identity,

$$E(T) = E(\sigma)E(X). \tag{3.7}$$

It follows that for a stable queuing system, $E(\sigma) < \infty$, that is, the mean number of services in a renewal cycle is finite.

Arrival and service rates, and traffic intensity: Let $\mu_A = E(X)$ be the mean inter-arrival time of customers. The arrival rate of the system is defined to be $a = 1/\mu_A$. This is the long-run average number of arrivals per unit time because by the SLLN,

$$a = \frac{1}{E(X)} = \lim_{n \to \infty} \frac{n}{X_1 + \cdots + X_n} = \lim_{t \to \infty} \frac{\# \text{ of arrivals by time } t}{t}.$$

Suppose the system has k servers. Let μ_i be the mean service time of server i. His service rate is defined to be $b_i = 1/\mu_i$. The (total) service rate of the system is defined to be $b = b_1 + b_2 + \cdots + b_k$. This is the long-run average number of services the system is capable to perform per unit time.

The traffic intensity is defined to be $\rho = a/b$. For k homogeneous servers,

$$\rho = E(Y)/[kE(X)],$$

where Y is a generic service time.

Busy and idle times for G/G/1: In a stable G/G/1 system, the length of a busy period is $B = \sum_{n=1}^{\sigma} Y_n$, where Y_n are the successive service times in a cycle. Note that Y_1 is the service time of the customer arriving at time 0 and Y_2 is the service time of the customer arriving at time X_1. By Wald's identity,

$$E(B) = E(\sigma)E(Y). \tag{3.8}$$

By considering a renewal reward process based on renewal cycles of the queuing system and with reward earned at unit rate whenever the system is busy, by (3.7) and (3.8), we obtain the following expression for the long-run fraction of busy time:

$$\lim_{t \to \infty} \frac{\text{amount of busy time by time } t}{t} = \frac{E(B)}{E(T)} = \rho. \tag{3.9}$$

Thus, $\rho \leq 1$ in a stable G/G/1 system, and the long-run fraction of idle time is $1 - \rho$.

Theorem 3.11 (*stability of G/G/1*) *A G/G/1 queuing system is stable if $\rho < 1$ (that is, if $a < b$). On the other hand, if $\rho > 1$, then almost surely the queue length converges to infinity, and consequently the system is not stable.*

Proof: We may assume that the system starts at time 0 with a single customer at the beginning of his service. As before, let X_n, Y_n, and T_n be, respectively, the successive inter-arrival times, service times and renewal cycle times. By SLLN, $(1/n) \sum_{k=1}^{n} X_k \to E(X)$ and $(1/n) \sum_{k=1}^{n} Y_k \to E(Y)$ a.s. as $n \to \infty$. If $a > b$, then $E(X) < E(Y)$ and hence $\sum_{k=1}^{n} X_k < \sum_{k=1}^{n} Y_k$ for large n. Therefore, for all large n, the nth arrival will find the server busy and the queue will never become empty. To show the queue length will converge to infinity, choose a constant θ with $0 < \theta < 1$ and $E(X) < \theta E(Y)$. By the SLLN, $(1/n) \sum_{k=1}^{[\theta n]} Y_k \to \theta E(Y)$, where $[r]$ denotes the integer part of a real number r. It follows that $\sum_{k=1}^{n} X_k < \sum_{k=1}^{[\theta n]} Y_k$. This means that when the nth customer arrives, the $[\theta n]$th service has not been completed and there are at least $n - [\theta n]$ customers in the system. As $n - [\theta n] \to \infty$ as $n \to \infty$, this shows that the queue length goes to infinity.

On the other hand, if $a < b$, then $E(Y) < E(X)$, and by SLLN, $\sum_{k=1}^{n} X_k > \sum_{k=1}^{n} Y_k$ for all large n. This means that the service will eventually catch up with arrivals and hence the cycle time T is finite. In order to show $E(T) < \infty$, we need the following simple technical result.

Lemma 3.12 *Let Z_n be iid with $E(Z) > 0$ and let τ be the smallest index $n \geq 1$ such that $S_n = Z_1 + Z_2 + \cdots + Z_n > 0$, setting $\tau = \infty$ if all $S_n \leq 0$. Then $E(\tau) < \infty$.*

Note that if c is a sufficiently large positive constant, then $E(Z_n \wedge c) > 0$ and if τ^c is defined for the iid sequence $Z_n \wedge c$ as τ is defined for Z_n, then $\tau \leq \tau^c$. Thus, we may assume $Z_n \leq c$ for some $c > 0$. Fix an integer $k > 0$ and let $S_k^\tau = \sum_{n=1}^{\tau \wedge k} Z_n$. Because $Z_n \leq c$, it is easy to see that $S_k^\tau \leq c$. By Wald's identity, $E(S_k^\tau) = E(\tau \wedge k) E(Z)$. Thus, $E(\tau \wedge k) \leq c/E(Z)$ and $E(\tau) = \lim_k E(\tau \wedge k) \leq c/E(Z) < \infty$. The lemma is proved.

We now return to the proof of Theorem 3.11. Let $Z_n = X_n - Y_n$ and let τ be defined as in Lemma 3.12. Then $E(\tau) < \infty$. Note that τ is the number of customers served in a cycle and $T = \sum_{n=1}^{\tau} X_n$. By Wald's identity, $E(T) = E(\tau)E(X) < \infty$. $\diamond$

Note on $\rho = 1$: It can be shown (see Propositions 1.3 and 3.1 in [1, chapter X]) that for G/G/1, if $\rho = 1$, then $T < \infty$ a.s. but $E(T) = \infty$, and hence the system is not stable.

Note on multi-servers: For a G/G/k system with homogeneous servers, let X and Y be, respectively, generic inter-arrival and service times. It can be shown (see Corollary 2.5 in [1, chapter XII]) that if $\rho < 1$ and $P(Y < X) > 0$, then the system is stable. Note that when $k = 1$, $\rho < 1$ implies $P(Y < X) > 0$, but for $k > 1$, the condition $\rho < 1$ alone is not enough to ensure the stability as defined here. However, under $\rho < 1$ alone, the system is stable in a more general sense as described earlier (see Theorem 2.2 in [1, chapter XII]).

It is not hard to show that for a general multi-server system, $\rho > 1$ implies that the queue length converges to ∞ a.s. and hence the system is not stable.

Simulation of a queuing system: One may write a MATLAB® program to simulate a queuing system. For example, consider a system of a single server with an exponential arrival rate $a = 4$ and a service time uniformly distributed on $[0, 0.4]$ so that the service rate is $b = 5$. Figure 3.1 shows the output of a MATLAB® program that simulates this system up to time $t = 50$, and plots the total queue length (the number of customers in the system) as a function of time, and displays the proportion of idle time. Note that the theoretical value for the long-run fraction of idle time $1 - a/b = 0.2$.

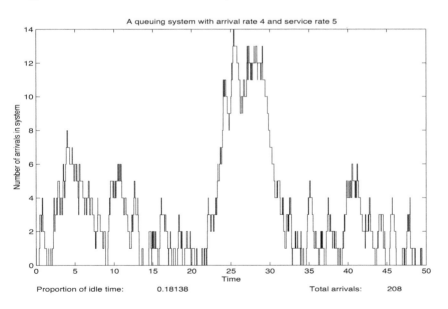

FIGURE 3.1: Total queue length of a queuing system.

3.4 Queue lengths, waiting times, and busy periods

Queue length processes: In a queuing system, let $Q(t)$ be the number of customers in the system at time t waiting for their service and let $\bar{Q}(t)$ be the total number in the system at time t either waiting or at service. We will call $Q(t)$ the queue length and $\bar{Q}(t)$ the total queue length processes. We will include in $Q(t)$ and $\bar{Q}(t)$ any customer who enters the system at time t, but not the one who leaves at t. Then $Q(t)$ and $\bar{Q}(t)$ are right continuous processes.

Proposition 3.13 *For a stable queuing system, the long-run time average Q of the queue length and the long-run time average $\bar{Q}$ of the total queue length, defined as*

$$Q = \lim_{t\to\infty} \frac{1}{t} \int_0^t Q(s)ds \quad \text{and} \quad \bar{Q} = \lim_{t\to\infty} \frac{1}{t} \int_0^t \bar{Q}(s)ds,$$

exist and are non-random, and are given by

$$Q = \frac{E \int_0^T Q(s)ds}{E(T)} \quad \text{and} \quad \bar{Q} = \frac{E \int_0^T \bar{Q}(s)ds}{E(T)}, \qquad (3.10)$$

where T is the length of a generic renewal cycle of the queuing system.

Proof: The result follows from the limiting properties of a cumulative renewal reward process (Theorem 3.6) by considering a reward earned at unit rate for each customer waiting for service or being in the system. $\diamondsuit$

Waiting times: Let W_n be the waiting time of the nth customer before his service and let $\bar{W}_n$ be the total time he spends in the system (including service time). The waiting time W_n is also called the actual waiting time to emphasize its difference from $\bar{W}_n$. We have $\bar{W}_n = W_n + Y_n$, where Y_n is the service time of nth customer. Note that the first customer arrives at time 0 when all servers are available, so $W_1 = 0$ and $\bar{W}_1 = Y_1$ is the first service time.

Proposition 3.14 *For a stable queuing system, the long-run average waiting time W and the long-run average total waiting time $\bar{W}$, defined by*

$$W = \lim_{n\to\infty} \frac{W_1 + \cdots + W_n}{n} \quad \text{and} \quad \bar{W} = \lim_{n\to\infty} \frac{\bar{W}_1 + \cdots + \bar{W}_n}{n},$$

exist and are non-random, and are given by

$$W = \frac{E(\sum_{i=1}^{\sigma} W_i)}{E(\sigma)} \quad \text{and} \quad \bar{W} = \frac{E(\sum_{i=1}^{\sigma} \bar{W}_i)}{E(\sigma)}, \tag{3.11}$$

where σ is the number of customers served in a renewal cycle of the queuing system. Moreover, for homogeneous servers, $\bar{W} = W + E(Y)$, where Y is a generic service time.

Proof: This can be proved by considering a renewal reward process, based on iid numbers σ_n of services in successive renewal cycles, with a reward W_i or $\bar{W}_i$ earned in the ith unit of time. In other words, each arriving customer advances the clock one unit time and generates a reward equal in value to his waiting time or total waiting time. $\diamond$

Theorem 3.15 *(Little's formula) In a stable queuing system with arrival rate a,*

$$Q = aW \quad \text{and} \quad \bar{Q} = a\bar{W}. \tag{3.12}$$

Proof: It is easy to see that $\int_0^T Q(s)ds = \sum_{i=1}^{\sigma} W_i$ (Imagine that each customer pays \$1 per unit time while waiting for service). By (3.7), the first equality in (3.12) follows from the first equalities in (3.10) and (3.11). The second equality is proved in the same way. $\diamond$

Theorem 3.16 *In a stable M/G/1 queuing system with arrival rate a, service rate b, and traffic intensity $\rho = a/b$ (< 1), let T and B be, respectively, the lengths of a renewal cycle and a busy period, and let σ be the number of services in a cycle. Then*

$$E(B) = \frac{1}{b-a}, \quad E(T) = \frac{1}{a} + \frac{1}{b-a} = \frac{1}{a(1-\rho)},$$

$$E(\sigma) = \frac{b}{b-a} = \frac{1}{1-\rho}. \tag{3.13}$$

Proof: Suppose the formula for $E(B)$ is proved. Since a cycle is a busy period followed by an idle period, and by the lack of memory property of exponential inter-arrivals, the length of the idle period has the same distribution as an inter-arrival time of mean $1/a$. This implies the formula for $E(T)$. Note that the use of the lack of memory property here is intuitive, but not rigorous. A more formal proof is to note that the time at the end of the first busy period is a stopping time of the

Poisson process $N_A(t)$ of arrivals, by the time homogeneity of $N_A(t)$, the length of the idle period has the same distribution as an inter-arrival time X.

The formula for $E(\sigma)$ now follows from (3.7). It remains to prove the formula for $E(B)$.

We may assume that at time 0, a customer arrives to start a busy period. During his service time Y_1, $N_A(Y_1)$ customers arrive. These customers are listed as $c_1, c_2, c_3 \ldots$. It is easy to see that as long as B is concerned, the customers can be served in any order. Thus, after the first service Y_1, the server can serve c_1 and all those who arrive during c_1's service, call c_1's descendants, as well as descendants of c_1's descendants, etc., until all customers descending from c_1 are served. Then the server will serve c_2 and all his descendants, c_3 and all his descendants, until the system becomes empty. This is called the bullpen discipline. We can then deduce

$$B = Y_1 + \sum_{i=1}^{N_A(Y_1)} B_i,$$

where B_i are iid, identical in distribution as B and independent of $N_A(Y_1)$. Then $E(B) = E(Y) + E[N_A(Y)]E(B)$. Solve for $E(B)$ to get $E(B) = E(Y)/\{1 - E[N_A(Y)]\}$. Because

$$E[N_A(Y)] = E\{E[N_A(Y) \mid Y]\} = E(aY) = aE(Y) = a/b,$$

it follows that $E(B) = (1/b)/[1 - (a/b)] = 1/(b - a)$. $\Diamond$

Theorem 3.17 *In a stable M/G/1 queue, the Q and W in Little's formula (3.12) are finite if and only if $E(Y^2) < \infty$, where Y is a generic service time. In this case,*

$$Q = \frac{a^2 E(Y^2)}{2(1 - \rho)} = \frac{\rho^2}{2(1 - \rho)} + \frac{a^2}{2(1 - \rho)} \mathrm{Var}(Y),$$

$$\bar{Q} = \rho + \frac{a^2 E(Y^2)}{2(1 - \rho)}. \tag{3.14}$$

The second equality for Q above means that an increase in the variance of service time causes a proportional increase in the system congestion (in terms of average queue length Q).

Proof: We will present a proof based on the bullpen discipline here. For a shorter proof but requiring more preparation, see Exercise 3.26. Let $Q_T = \int_0^T Q(s)ds$. Let Y_1 be (the length of) the initial service period. By the bullpen discipline in the proof of Theorem 3.16,

$$Q_T = H_1 + H_2 + \sum_{i=1}^{N_A(Y_1)} Q_{T,i},$$

where $H_1 = \sum_{i=1}^{N_A(Y_1)} (Y_1 - S_i)$ (S_i is the arrival time of the ith customer) is the amount of time spent in the system during Y_1 by those customers arriving during Y_1,

$$H_2 = (n-1)B_1 + (n-2)B_2 + \cdots + B_{n-1},$$

with $n = N_A(Y_1)$, and B_i being the successive busy periods initiated by the customers arriving during Y_1, is the amount of waiting time by those customers between the end of Y_1 and the beginning of their services, and $Q_{T,i}$ are iid, identical in distribution as Q_T and independent of $N_A(Y_1)$. Taking expectation, we get

$$E(Q_T) = E(H_1) + E(H_2) + E[N_A(Y)]E(Q_T).$$

By (2.4), $E[\sum_{i=1}^{N_A(t)} (t - S_i)] = at^2/2$ and hence $E(H_1) = E[E(H_1|Y)] = E(aY^2/2) = (a/2)E(Y^2)$.

$$E(H_2) = E\{E[H_2 \mid N_A(Y_1)]\} = E\{\tfrac{1}{2}N_A(Y)[N_A(Y) - 1]\}E(B)$$

$$= \tfrac{1}{2}E[(aY)^2]E(B) = \frac{a^2}{2(b-a)}E(Y^2).$$

Since $E[N_A(Y)] = a/b$,

$$E(Q_T) = [\frac{a}{2} + \frac{a^2}{2(b-a)}]E(Y^2) + \frac{a}{b}E(Q_T).$$

If $E(Q_T)$ is known to be finite, then it may be solved from the above equation to get

$$E(Q_T) = \frac{ab^2}{2(b-a)^2}E(Y^2)$$

and

$$Q = \frac{E(Q_T)}{E(T)} = \frac{E(Q_T)}{b/[a(b-a)]} = \frac{a^2b}{2(b-a)}E(Y^2) = \frac{a^2 E(Y^2)}{2(1-\rho)}.$$

This proves the relation for Q in (3.14), which also implies the relation for $\bar{Q}$ as $\bar{Q} = a\bar{W} = aW + a/b = Q + \rho$. It also shows that if Q is finite, then $E(Y^2) < \infty$.

It remains to show that if $E(Y^2) < \infty$, then $E(Q_T) < \infty$. For any integer $c > 0$, let $Q_T^c = \int_0^T [Q(s) \wedge c] ds$. This is the total time spent in the system by customers served in the first busy period, but when the number of customers exceeds c, the time will only be counted for c customers. Then the bullpen discipline will lead to the inequality

$$Q_T^c \le H_1 + H_2 + \sum_{i=1}^{N_A(Y_1)} Q_{T,i}^c$$

with $Q_{T,i}^c$ being iid, identical in distribution to Q_T^c and independent of $N_A(Y_1)$. Taking expectation,

$$E(Q_T^c) \le E(H_1) + E(H_2) + E[N_A(Y)]E(Q_T^c).$$

Solve for $E(Q_T^c)$ to get

$$E(Q_T^c) \le \frac{E(H_1) + E(H_2)}{1 - E[N_A(Y)]} = \frac{ab^2}{2(b-a)^2} E(Y^2)$$

and

$$E(Q_T) = \lim_{c \uparrow \infty} E(Q_T^c) \le \frac{ab^2}{2(b-a)^2} E(Y^2) < \infty. \quad \diamondsuit$$

Note: A waiting time W_n is contained in a cycle of the queuing system, and so is less than the cycle length, but in an M/G/1 system, the long-run average waiting time $W = Q/a = aE(Y^2)/[2(1-\rho)]$ can be greater than the mean cycle length $E(T) = 1/[a(1-\rho)]$. In fact, W and Q can even be infinite in a stable M/G/1 queue; see Exercise 3.12 below. This seemingly paradoxical phenomenon is due to the fact that waiting times are overlapped.

Exercise 3.11 Customers arrive at a single server station according to a Poisson process of rate $\lambda = 3$ per unit time. Each customer, independent of others, requires a random service time uniformly distributed between 0 and 0.5. Find
(a) the long-run fraction of time when the server is idle, and
(b) the average amount of time a customer waits before service.

Exercise 3.12 Consider an M/G/1 system with mean exponential inter-arrival time 5 and a continuous service time with pdf $f(y) = 4/y^3$ for $y > 1$. Show the system is stable and find the expected cycle time $E(T)$. Also show the long-run average queue length Q and waiting time W are infinite.

Exercise 3.13 (waiting times in G/G/1) Consider a G/G/1 system.
(a) Show the waiting times (before service) satisfy the following recursive relation (recall $W_1 = 0$): For $n \geq 1$,

$$W_{n+1} = (W_n + Y_n - X_n)_+, \qquad (3.15)$$

where X_n is the time between nth and $(n+1)$st arrivals, W_n and Y_n are, respectively, the waiting time and the service time of nth customer, and $a_+ = \max(a, 0)$ for any real number a.
(b) Let $Z_n = Y_n - X_n$. Then Z_n are iid. Show that for $n \geq 1$,

$$W_{n+1} = \max\{0, Z_n, Z_n + Z_{n-1}, \ldots, Z_n + Z_{n-1} + \cdots + Z_1\}. \qquad (3.16)$$

(c) Let $S_n = Z_1 + \cdots + Z_n$ with $S_0 = 0$. Show that for $n \geq 1$

$$W_{n+1} \overset{d}{=} \max\{S_0, S_1, S_2, \ldots, S_n\} \quad \text{(equal in distribution)}. \qquad (3.17)$$

Moreover, if the G/G/1 system is stable, then $\max\{S_0, S_1, S_2, \ldots, S_n\} \uparrow V$ as $n \uparrow \infty$ for some finite random variable V, and hence $W_n \overset{d}{\to} V$.

Exercise 3.14 Consider a G/M/1 system with uniform inter-arrival time on $(0, 1)$ and exponential service time of mean 1. Assume the server is idle when the first customer arrives, that is, $W_1 = 0$. Find the probability that the waiting times of next two customers are both less than 1, that is, $P(W_2 \leq 1, W_3 \leq 1)$.
Hint: It may be helpful to get $P(W_2 \leq w)$ for any $w \geq 0$ first.

Exercise 3.15 (M/G/1 with delayed service) Consider a M/G/1 system with arrival rate a and service rate b such that $a < b$. Assume the server begins service only when there are $m \geq 1$ customers in the system, and once the service starts, it will continue until all customers are cleared from the system.
(a) Identify a sequence of renewal cycles associated with this system and show $E(T) = mE(T^1)$, where T is the length of a cycle and the superscript 1 indicates the corresponding quantity for the standard M/G/1 with $m = 1$.

(b) Verify that Proposition 3.13 for Q and $\bar{Q}$, Proposition 3.14 for W and $\bar{W}$, and Little's formula Theorem 3.15 hold for the delayed service.
(c) Show

$$Q = Q^1 + (m-1)/2 \quad \text{and} \quad \bar{Q} = \bar{Q}^1 + (m-1)/2.$$

Hint: Apply the bullpen discipline to the initial m customers to show $E(B) = mE(B^1)$, where B is the length of a busy period.

3.5 Renewal equation

Renewal equation: A function on $\mathbb{R}_+ = [0, \infty)$ is automatically assumed to be 0 on $(-\infty, 0)$. It is called locally bounded if it is bounded on $[0, b]$ for any $b > 0$. Let $V(t)$ be an increasing function and let $Z(t)$ be a locally bounded function on $\mathbb{R}_+$. The convolution $Z * V$ is defined as the convolution of two distribution functions given in §1.4, that is,

$$(Z * V)(t) = \int_{[0,\infty)} Z(t-s)dV(s) = \int_{[0,t]} Z(t-s)dV(s).$$

Note that $(Z * V)$ is also a locally bounded function on $\mathbb{R}_+$.

Given a renewal process $N(t)$ with cycle time distribution F and renewal function $m(t)$, and let $z(t)$ be an arbitrary function on $\mathbb{R}_+$. The following equation for an unknown function $Z(t)$ is called a renewal equation:

$$Z(t) = z(t) + (Z * F)(t). \tag{3.18}$$

Theorem 3.18 *Let $z(t)$ be a locally bounded function on $\mathbb{R}_+$. Then*

$$Z(t) = z(t) + (z * m)(t) \tag{3.19}$$

is the unique locally bounded solution to the renewal equation (3.18).

Proof: By (3.2), it is easy to show $Z(t) = z(t) + (z * m)(t)$ is indeed a solution of (3.18) and is locally bounded. Suppose $Z_2(t)$ is another locally bounded solution of (3.18), then $H(t) = Z(t) - Z_2(t)$ is a locally bounded solution of $H = H * F$. Then $H = H * F = H * F * F = \cdots = H * F^{*n}$ for any integer $n > 0$, and for any finite $b > 0$

$$\sup_{0 \le t \le b} |H(t)| \le \sup_{0 \le t \le b} |H(t)| F^{*n}(b) = \sup_{0 \le t \le b} |H(t)| P(S_n \le b) \to 0$$

as $n \to \infty$. $\diamond$

T_1-shift invariance: Let T_1 be the first renewal time of the renewal process $N(t)$. A process $X(t)$ is called T_1-shift invariant if for $x < t$, the conditional distribution of $X(t)$ given $T_1 = x$ is the same as the distribution of $X(t - x)$. In particular, this implies that if $X(t)$ is real valued, either bounded or ≥ 0, then

$$E[X(t) \mid T_1 = x] = E[X(t - x)] \quad \text{for } x < t. \tag{3.20}$$

For example, the queue length and the total queue length processes are T_1-shift invariant under the renewal process of a stable queuing system.

A time dependent event $G(t)$ is called T_1-shift invariant if its indicator $1_{G(t)}$ is so. This is equivalent to

$$P[G(t) \mid T_1 = x] = P[G(t - x)] \quad \text{for } x < t. \tag{3.21}$$

For example, $G(t)$ may be the event that all servers are free at time t in a stable queuing system.

Proposition 3.19 *A process $X(t)$ is T_1-shift invariant if and only if for any $t \geq 0$, $X(T_1 + t)$ is independent of T_1 and has the same distribution as $X(t)$.*

Proof: First assume $X(t)$ is T_1-shift invariant. Then for any bounded Borel function f and $x, t \geq 0$,

$$E[f(X(T_1 + t)) \mid T_1 = x] = E[f(X(x + t)) \mid T_1 = x] = E[f(X(t))],$$

and $E[f(X(T_1 + t))] = \int_{[0, \infty)} E[f(X(T_1 + t)) \mid T_1 = x] dF(x) = E[f(X(t))]$. This being true for any f and x shows that $X(T_1 + t)$ is independent of T_1 and has the same distribution as $X(t)$. Conversely, if this holds, then for any bounded Borel function f and $x < t$,

$$E[f(X(t)) \mid T_1 = x] = E[f(X(x + t - x)) \mid T_1 = x]$$
$$= E[f(X(T_1 + t - x)) \mid T_1 = x] = E[f(X(t - x))]. \quad \diamond$$

Conditioning on the first renewal: Let $X(t) \geq 0$ be T_1-shift invariant and let $Z(t) = E[X(t)]$. By the total probability law,

$$Z(t) = E[X(t); T_1 > t] + \int_{[0, t]} Z(t - x) dF(x). \tag{3.22}$$

This shows that $Z(t)$ satisfies the renewal equation (3.18) with $z(t) = E[X(t); T_1 > t]$, and hence by Theorem 3.18,

$$E[X(t)] = E[X(t); T_1 > t] + \int_{[0,t]} E[X(t-x); T_1 > t-x] dm(x). \quad (3.23)$$

Note that (3.23) holds even when $E[X(t); T_1 > t]$ is not locally bounded because it holds when $X(t)$ is replaced by $X(t) \wedge b$ for $b > 0$ and letting $b \to \infty$ shows that it holds in general.

For a T_1-shift invariant event $G(t)$, it follows from (3.23) that

$$P[G(t)] = P[G(t), T_1 > t] + \int_{[0,t]} P[G(t-x), T_1 > t-x] m(dx). \quad (3.24)$$

A note on measurability: In the above we have to assume the functions $E[X(t)]$ and $P[G(t)]$ are measurable in t, then so are $E[X(t); T_1 > t]$ and $P[G(t), T_1 > t]$ by (3.22), and hence they can be integrated. If $X(t)$ (resp. $1_{G(t)}$) is either right or left continuous, then it is measurable on the product space $\mathbb{R}_+ \times \Omega$, and then by Fubini Theorem in measure theory, $E[X(t)]$ (resp. $P[G(t)]$) is measurable in t.

Current age and excess life: The current age is $A(t) = t - S_{N(t)}$, the length of time between the last renewal and present time t, and the excess life is $B(t) = S_{N(t)+1} - T$, the length of time between the present time t and the next renewal. They are also called, respectively, the backward and forward recurrence times.

Note that both $A(t)$ and $B(t)$ are T_1-shift invariant,

$$E[A(t); T_1 > t] = t\bar{F}(t)$$

and

$$
\begin{aligned}
E[B(t); T_1 > t] &= E[T_1 - t; T_1 > t] = E\left[\int_t^\infty 1_{[T_1 > x]} dx\right] \\
&= \int_t^\infty P(T_1 > x) dx = \int_t^\infty \bar{F}(x) dx.
\end{aligned}
$$

Then by (3.23),

$$E[A(t)] = t\bar{F}(t) + (t\bar{F}) * m(t) = t\bar{F}(t) + \int_{[0,t]} (t-s)\bar{F}(t-s) dm(s) \quad (3.25)$$

and

$$E[B(t)] = \int_t^\infty \bar{F}(x)dx + (\int_t^\infty \bar{F}(x)dx) * m(t) \qquad (3.26)$$

$$= \int_t^\infty \bar{F}(x)dx + \int_{[0,\,t]} [\int_{t-s}^\infty \bar{F}(x)dx]dm(s).$$

The distribution of $A(t)$ and $B(t)$ may be determined as follows. For $y \geq 0$, both events $[A(t) \leq y]$ and $[B(t) > y]$ are T_1-shift invariant with

$$P[A(t) \leq y, T_1 > t] = \bar{F}(t)1_{[0,\,y]}(t)$$

and

$$P[B(t) > y, T_1 > t] = \bar{F}(t+y),$$

and hence by (3.24),

$$P[A(t) \leq y] = \bar{F}(t)1_{[0,\,y]}(t) + (\bar{F}1_{[0,\,y]} * m)(t) \qquad (3.27)$$

$$= \bar{F}(t)1_{[0,\,y]}(t) + \int_{[(t-y)\vee 0,\,t]} \bar{F}(t-s)dm(s)$$

$$(a \vee b = \max(a, b)),$$

$$P[B(t) > y] = \bar{F}(t+y) + (\bar{F}(t+y) * m)(t) \qquad (3.28)$$

$$= \bar{F}(t+y) + \int_{[0,\,t]} \bar{F}(t+y-s)dm(s).$$

Length of renewal cycle straddling time t: The length of the renewal cycle straddling time t is $T(t) = A(t) + B(t)$. For a Poisson process, by the lack of memory property of the exponential cycle times, the excess life $B(t)$ and a generic cycle length T have the same distribution, and hence $E[T(t)] = E[A(t)] + E(T) > E(T)$ for $t > 0$. Thus, the cycle straddling t is longer than a typical cycle in average. By Theorem 3.20 and Exercise 3.19 below, this fact is true in general and is called the inspector's paradox in the sense that the inspected renewal cycles tend to be larger than a typical cycle.

Theorem 3.20 *For all $t, x \geq 0$, $\bar{F}_{T(t)}(x) \geq \bar{F}(x)$ and hence*

$$E[T(t)] \geq E(T).$$

Proof: For $t, y \geq 0$, we compute,

$$P[T(t) > y]$$

$$= P(T_1 > y, T_1 > t) + \sum_{n=1}^\infty P(T_{n+1} > y, S_n \leq t, S_{n+1} > t)$$

$$
\begin{aligned}
&= \ P(T_1 > y, T_1 > t) + \sum_{n=1}^{\infty} \int_{[0,\,t]} P(T_{n+1} > y, T_{n+1} > t - u) dF_n(u) \\
&= \ P(T > y \mid T > t) P(T > t) \\
&\quad + \sum_{n=1}^{\infty} \int_{[0,\,t]} P(T > y \mid T > t - u) P(T > t - u) dF_n(u).
\end{aligned}
$$

If $y \le t$, then $P(T > y \mid T > t) = 1$, which is $\ge P(T > y)$, and if $y > t$, then $P(T > y \mid T > t) = P(T > y)/P(T > t) \ge P(T > y)$. Thus,

$$
\begin{aligned}
&P[T(t) > y] \\
&\ge \ P(T > y)[P(T > t) + \sum_{n=1}^{\infty} \int_{[0,\,t]} P(T > t - u) dF_n(u)] \\
&= \ P(T > y)[P(T_1 > t) + \sum_{n=1}^{\infty} P(S_n \le t, S_n + T_{n+1} > t)] \\
&= \ P(T > y).
\end{aligned}
$$

and

$$
E[T(t)] = \int_0^{\infty} P[T(t) > y] dy \ge \int_0^{\infty} P(T > y) dy = E(T). \quad \Diamond
$$

Example 3.21 For a Poisson process of rate λ, for any t and $y > 0$, find $P[T(t) > y]$.

Solution: The event $Z(t) = [T(t) > y]$ is T_1-shift invariant, so it satisfies the renewal equation (3.18) with $z(t) = P[T(t) > y, T_1 > t] = P[T_1 > y, T_1 > t] = \bar{F}(y \vee t) = e^{-\lambda(y \vee t)}$. By Theorem 3.18,

$$
\begin{aligned}
P[T(t) > y] &= \ e^{-\lambda(y \vee t)} + \int_0^t e^{-\lambda(y \vee (t-s))} \lambda ds \\
&= \ e^{-\lambda(y \vee t)} + \int_0^t e^{-\lambda(y \vee s)} \lambda ds = [1 + \lambda(y \wedge t)] e^{-\lambda y}.
\end{aligned}
$$

Exercise 3.16 For a Poisson process $N(t)$ of rate λ, determine the distribution of the current age $A(t)$ and the excessive life $B(t)$, and also find their means $E[A(t)]$ and $E[B(t)]$.

Exercise 3.17 Shocks occur to a system according to a renewal process $N(t)$ with cycle time distribution $F(t)$. Each shock causes the

system to undergo repair which lasts a random time Y of distribution $F_Y(t)$. A shock occurring during a repair will cause the repair to be aborted and a new repair starts immediately. Assume shocks and repairs are independent, and a shock occurs at time 0.
(a) Derive a renewal equation for $f(t) = P(\text{system is under repair at time } t)$.
(b) Solve the renewal equation assuming $N(t)$ is a Poisson process of rate λ and Y is uniformly distributed on $(0, 1)$.

Exercise 3.18 Let $Z(t) = P[A(t) \le x, B(t) \le y]$ for $x, y \ge 0$.
(a) Show that $f(t)$ satisfies a renewal equation.
(b) Solve the renewal equation for a Poisson process of rate λ to find the joint distribution function $F(x, y) = P[A(t) \le x, B(t) \le y]$.
(c) Find $P[A(t) > B(t)]$ for a Poisson process of rate λ.
Hint: The probability in (c) may be found either by solving a renewal equation or by a direct computation using $F(x, y)$ from (b).

Exercise 3.19 (a) Show that for any $t > 0$, if $E[T(t)] = E(T)$, then $T(t)$ and T have the same distribution.
(b) Assume F has a continuous distribution with pdf $f(t) > 0$ for $t > 0$. Show that for any $t > 0$, $E[T(t)] > E(T)$.

3.6 Key renewal theorem

Direct Riemann integration: Let $z(t)$ be a nonnegative function on $\mathbb{R}_+$. For any real number $h > 0$, define

$$\underline{\sigma}(h) = \sum_{k=1}^{\infty} h\left[\inf_{(k-1)h \le t < kh} z(t)\right] \quad \text{and} \quad \overline{\sigma}(h) = \sum_{k=1}^{\infty} h\left[\sup_{(k-1)h \le t < kh} z(t)\right].$$

Note that $\underline{\sigma}(h)$ and $\overline{\sigma}(h)$ are, respectively, the areas below the graphs of two step functions which sandwich the graph of $z(t)$ in between. It can be shown that $\underline{\sigma}(h) \le \overline{\sigma}(h')$ for any $h, h' > 0$, and $\underline{\sigma}(h) \uparrow$ and $\overline{\sigma}(h) \downarrow$ as $h \downarrow 0$. The function $z(t)$ will be called directly Riemann integrable, or dRi in short, if $\underline{\sigma}(h)$ and $\overline{\sigma}(h)$ converge to a common finite limit as $h \to 0$. In this case, the direct Riemann integral $\int_0^{\infty} z(t)dt$ of $z(t)$ is defined to be the common limit $\lim_{h \to 0} \overline{\sigma}(h) = \lim_{h \to 0} \underline{\sigma}(h)$.

By the definition of the classical Riemann integrals, if $z(t)$ is dRi, then it is integrable over $[0, \infty)$ as an improper Riemann integral with the same integral value. It is easy to show that for a nonnegative and decreasing function $z(t)$ on $\mathbb{R}_+$, if $\int_0^\infty z(t)dt$ exists and is finite as an improper Riemann integral, then $z(t)$ is dRi with same integral value. Moreover, if $z(t) \geq 0$ is Riemann integrable on any finite interval and if $z(t) \leq g(t)$ for all $t \geq 0$ for some dRi $g(t)$, then $z(t)$ is also dRi. See [9, Subsection 3.10.1] for more details.

Lattice distribution: A random variable X is said to have a lattice distribution if X only takes values that are multiples of some $h > 0$, that is, if $P\{\cup_{n=-\infty}^\infty [X = hn]\} = 1$. The largest $h > 0$ for this property is called the span of X or of its distribution.

A random variable or its distribution is called nonlattice if it does not have a lattice distribution. A stable queuing system is called non-lattice if its renewal cycle time is so. This is the case if and only if the inter-arrival time distribution is nonlattice (see Proposition 3.2 in [1, chapter X]).

Theorem 3.22 *(Key renewal theorem) Assume the cycle time distribution F is non-lattice of mean μ. If $z(t)$ is dRi, then*

$$\lim_{t\to\infty} z * m(t) = \mu^{-1} \int_0^\infty z(t)dt. \qquad (3.29)$$

For a lattice F with span h, the result takes the following form: If $z(t)$ is a nonnegative function on $\mathbb{R}_+$ satisfying $\sum_{n=0}^\infty z(t + hn) < \infty$ for some t with $0 \leq t < h$, then

$$\lim_{n\to\infty} z * m(t + hn) = \mu^{-1}h \sum_{n=0}^\infty z(t + hn). \qquad (3.30)$$

In (3.29) and (3.30), the mean cycle time μ may be infinite.

For the proof of the key renewal theorems, see [9, section 3.10] or [11, section 2.7].

Limiting expectation and probability: Let the cycle time distribution F be non-lattice with a finite mean μ. Let $X(t) \geq 0$ be a bounded T_1-shift invariant process as defined in §3.5. Applying the key renewal theorem to (3.23), we obtain

$$\lim_{t\to\infty} E[X(t)] = \frac{1}{\mu} \int_0^\infty E[X(t); T_1 > t]dt. \qquad (3.31)$$

In particular, if $G(t)$ is a T_1-shift invariant event, then

$$\lim_{t \to \infty} P[G(t)] = \frac{1}{\mu} \int_0^\infty P[G(t), T_1 > t] dt = \frac{1}{E(T_1)} E\Big[\int_0^{T_1} 1_{G(t)} dt\Big]$$

$$= \frac{E[\text{amount time in a cycle when event occurs}]}{E[\text{length of a cycle}]}. \tag{3.32}$$

Checking dRi: In (3.31), one should assume $z(t) = E[X(t); T_1 > t]$ is dRi. This holds if $X(t) \geq 0$ is bounded, and is either left or right continuous, because then so is $Z(t) = E[X(t)]$, and by the renewal equation, $z(t)$ has at most countably many discontinuous points. This implies that $z(t)$ is Riemann integrable. Because $z(t) \leq \bar{F}(t)$ and $\bar{F}(t)$ is dRi, it follows that $z(t)$ is dRi. Similarly, $z(t) = P[G(t), T_1 > t]$ in (3.32) should be dRi, which holds if $1_{G(t)}$ is either left or right continuous.

Example 3.23 Consider the on-off system in Example 3.7. The length of a on-off cycle is $T = X + Y$. Then $G(t) = [\text{system is on at time } t]$ is a T_1-shift invariant event, so $Z(t) = P[G(t)]$ satisfies the renewal equation (3.18) with $z(t) = P[G(t), T_1 > t] = P(X > t) = \bar{F}_X(t)$. By Theorem 3.18, $Z(t) = z(t) + z * m(t)$ and by the key renewal theorem,

$$\lim_{t \to \infty} P[\text{on at time } t] = \frac{\int_0^\infty \bar{F}_X(t) dt}{E(T)} = \frac{E(X)}{E(X) + E(Y)}.$$

This is just the long-run fraction of on time obtained in Example 3.7 using a renewal reward process, which is no coincidence because by (3.32), the limiting probability of a T_1-shift invariant event has the same form as the long-run fraction of time when the event occurs. In fact, one may obtain the limiting probability more quickly using (3.32). However, (3.32) is a consequence of the key renewal theorem, not that of a much simpler theory of renewal reward processes.

Theorem 3.24 *(Limiting distributions of $A(t)$ and $B(t)$)* *Assume the cycle time T has a nonlattice distribution F with a finite mean μ. Then the current age $A(t)$ and excessive life $B(t)$ have the same limiting distribution given by*

$$\lim_{t \to \infty} P[A(t) \leq x] = \lim_{t \to \infty} P[B(t) \leq x] = \frac{1}{\mu} \int_0^x \bar{F}(t) dt \tag{3.33}$$

for $x \geq 0$. Moreover, the limiting means of $A(t)$ and $B(t)$ are also equal and are given by

$$\lim_{t \to \infty} E[A(t)] = \lim_{t \to \infty} E[B(t)] = \frac{1}{\mu} \int_0^\infty t\bar{F}(t) dt = \frac{E(T^2)}{2E(T)}. \tag{3.34}$$

Note: The limiting mean (3.34) is the mean of the limiting distribution (3.33), which is to be expected, but may not be true in general.

Proof: By (3.27), (3.28), and (3.29), $\lim_{t\to\infty} P[A(t) \le y] = (1/\mu) \int_0^y \bar{F}(s)ds$ and

$$\lim_{t\to\infty} P[B(t) > y] = (1/\mu) \int_y^\infty \bar{F}(s)ds = 1 - (1/\mu) \int_0^y \bar{F}(s)ds,$$

noting that both $\bar{F}(t)1_{[0,y]}(t)$ and $\bar{F}(t+y)$ are dRi as functions of t, and $\mu = \int_0^\infty \bar{F}(s)ds$.

Let $z_A(t) = E[A(t); T_1 > t]$ and $z_B(t) = E[B(t); T_1 > t]$. As shown in §3.5, $z_A(t) = t\bar{F}(t)$ and $z_B(t) = \int_t^\infty \bar{F}(x)dx$. Then $\int_0^\infty z_A(t)dt = E[\int_0^\infty t1_{[T>t]}dt] = E[\int_0^T tdt] = (1/2)E(T^2)$ and a change of integration order shows that $\int_0^\infty z_B(t)dt = \int_0^\infty x\bar{F}(x)dx = \int_0^\infty z_A(t)dt$. Moreover, both $z_A(t)$ and $z_B(t)$ are bounded by $E(T; T > t)$, which converges to 0 as $t \to \infty$, and $\int_0^\infty E(T; T > t)dt = E(T \int_0^T dt) = E(T^2)$. It follows that if $E(T^2) < \infty$, then both $z_A(t)$ and $z_B(t)$ are dRi, and (3.34) is implied by (3.25), (3.26) and (3.29).

Now suppose $E(T^2) = \infty$. Fix $b > 0$. Similar to the derivation of the renewal equations (3.25) and (3.26) for $E[A(t)]$ and $E[B(t)]$, it is easy to show that $Z(t) = E[A(t) \wedge b]$ or $Z(t) = E[B(t) \wedge b]$ satisfies the renewal equation (3.18) with $z(t) = (t \wedge b)\bar{F}(t)$ or $z(t) = E[(T-t) \wedge b; T > t]$. By (3.29), $\lim_{t\to\infty} E[Z(t)] = (1/\mu) \int_0^\infty z(t)dt$. In both cases, $\int_0^\infty z(t)dt \uparrow E(T^2)/2 = \infty$ as $b \uparrow \infty$ by the computation in the last paragraph. Because $E[A(t)] \ge E[A(t) \wedge b]$ and $\lim_{t\to\infty} E[A(t) \wedge b] \uparrow \infty$, it follows that $\lim_{t\to\infty} E[A(t)] = \infty$. Similarly, $\lim_{t\to\infty} E[B(t)] = \infty$. This proves (3.34) completely. $\diamondsuit$

Stationary renewal process: A delayed renewal process $N_D(t)$ is said to be a stationary renewal process if for any constant $u > 0$, the time-shifted counting process $N_D^u(t) = N_D(u + t) - N_D(u)$ has the same distribution as the process $N_D(t)$. If we can assume the limiting excessive life distribution (3.33) still holds for a delayed renewal process, then the excessive life distribution of a stationary renewal process $N_D(t)$ at any time t should be given by (3.33). In particular, the initial renewal time distribution G (that is, the distribution of T_1) should be given by

$$G(x) = \frac{1}{\mu} \int_0^x \bar{F}(t)dt, \qquad x \ge 0. \tag{3.35}$$

In fact, by Theorem 3.25 below, a delayed renewal process $N_D(t)$ is a stationary renewal process if and only if its initial renewal time distribution G is given by (3.35).

Note that a stationary renewal process has stationary (but not necessarily independent) increments. By the limiting excessive life distribution (3.33), if one starts counting events at a large time $u > 0$ of a renewal process, then one obtains a stationary renewal process.

It is easy to see that a Poisson process is a stationary renewal process that is trivially delayed (meaning not delayed).

Theorem 3.25 *Let $N_D(t)$ be a delayed renewal process with cycle time distribution F and finite mean μ. Then $N_D(t)$ is a stationary renewal process if and only if its initial renewal time distribution G is given by (3.35). In this case, its mean function is $m_D(t) = t/\mu$.*

Proof: Assume (3.35). We first show $m_D(t) = t/\mu$. The Laplace transform of $m_D(t)$ is

$$\hat{m}_D = \sum_{n=0}^{\infty} (G * F^{*n})\hat{\ } = \sum_{n=0}^{\infty} \hat{G}\hat{F}^n = \frac{\hat{G}}{1 - \hat{F}}.$$

and the Laplace transform of G given by (3.35) is

$$\hat{G}(s) = \int_0^{\infty} e^{-st} dG(t) = \int_0^{\infty} e^{-st} \bar{F}(t)\frac{dt}{\mu} = \int_0^{\infty} e^{-st} \int_t^{\infty} dF(y)\frac{dt}{\mu}$$

$$= \int_0^{\infty} \int_0^y e^{-st}\frac{dt}{\mu} dF(y) = \frac{1}{\mu s} \int_0^{\infty} (1 - e^{-sy}) dF(y) = \frac{1 - \hat{F}(s)}{\mu s}.$$

It follows that $\hat{m}_D(s) = 1/(\mu s)$, which is the Laplace transform of the distribution type function $F(t) = t/\mu$, and hence $m_D(t) = t/\mu$.

As the distribution of a delayed renewal process is determined by its cycle time and the initial renewal time distributions, F and G, to show that $N_D(t)$ is a stationary renewal process, it is enough to show that all the excessive life distributions are equal to G. For $y \geq 0$,

$$P[B(t) > y]$$

$$= P[B(t) > y, T_1 > t] + \sum_{n=1}^{\infty} P[B(t) > y, S_n \leq t < S_{n+1}]$$

$$= P(T_1 - t > y, T_1 > t)$$

$$+ \sum_{n=1}^{\infty} \int_{[0, t]} P(u + T_{n+1} - t > y, u + T_{n+1} > t) dF_n(u)$$

$$= \bar{G}(t+y) + \sum_{n=1}^{\infty} \int_{[0,\,t]} \bar{F}(t+y-u)dF_n(u)$$

$$= \bar{G}(t+y) + \int_{[0,\,t]} \bar{F}(t+y-u)dm_D(u)$$

$$= \bar{G}(t+y) + \int_0^t \bar{F}(t+y-u)du/\mu$$

$$= \bar{G}(t+y) + \int_y^{t+y} \bar{F}(u)du/\mu = \bar{G}(y)$$

by (3.35). This proves that $N_D(t)$ is a stationary renewal process.

Now assume $N_D(t)$ is a stationary renewal process. Then $f(t) = E[N_D(t)]$ is a right continuous function satisfying $f(s+t) = f(t)+f(s)$. This implies $E[N_D(t)] = f(t) = at$ for some constant $a > 0$. As a stationary renewal process, $\bar{G}(y) = P(T_1 > y) = P[B(t) > y]$, which may be computed as above to get $\bar{G}(y) = \bar{G}(t + y) + a \int_y^{t+y} \bar{F}(u)du$. Letting $t \to \infty$ yields $\bar{G}(y) = a \int_y^\infty \bar{F}(u)du$. This implies $G(x)$ is given by (3.35) and $a = 1/\mu$. ◇

Exercise 3.20 Find the limiting expectation and distribution of the excessive life $B(t)$ for each of the following cycle time distributions.
(a) Uniform on $(0, 1)$.
(b) Erlang distribution $\text{Erlang}(n, \lambda)$.

Exercise 3.21 People arrive at a bus station according to a Poisson process at a rate of 1 every 2 minutes. A bus departs the station every 4 minutes or whenever there are 3 people waiting. Find the expected waiting time for the bus in the long run, that is,

$$\lim_{t\to\infty} E[\text{waiting time of a passenger arriving at time } t].$$

Exercise 3.22 Find the limit of $E[B(n)]$ as integer $n \to \infty$ for a geometric cycle time distribution with success probability p ($0 < p < 1$). Also show the limit of $E[B(t)]$ as real $t \to \infty$ does not exist. This is an example of lattice cycle time distribution.

Exercise 3.23 An item with exponential lifetime of mean $1/\lambda$ is to be replaced by an identical new item when it fails, but the replacement will take a random amount time of mean ν to complete. Let $f_x(t)$ be the probability that an item is working at time t and it will stay working for at least x more units of time. Find $\lim_{t\to\infty} f_x(t)$.

3.7 Regenerative processes

Definition: A continuous time process $X(t)$ with right continuous paths is called a regenerative process if there is a renewal sequence S_n such that $X(t)$ restarts probabilistically at each S_n. More precisely, this means that the time-shifted process $X_n(t) = X(S_n + t)$ has the same distribution as $X(t)$, and is independent of $\{S_1, S_2, \ldots, S_n\}$ and process $X(t)$ up to S_n (that is, the process $X(t)$ is independent of $\{S_1, S_2, \ldots, S_n\}$ and the stopped process $X^{\wedge S_n}(t) = X(t \wedge S_n)$). The S_n are called the associated renewal sequence and $T_n = S_n - S_{n-1}$ for $n \geq 1$ ($S_0 = 0$) the associated renewal cycle times.

By Proposition 3.19, a regenerative process is T_1-shift invariant.

As an example, the queue length $Q(t)$ or the total queue length $\bar{Q}(t)$ of a stable queuing system is a regenerative process that restarts at the beginning of each renewal cycle of the queuing system.

Theorem 3.26 *(Smith's theorem) Let $X(t)$ be a regenerative process based on iid renewal cycle times T_n. Assume it takes isolated discrete values (such as integers), and the distribution F of T_n is nonlattice with a finite mean μ. For any state j, let $q_j(t) = P[X(t) = j, T_1 > t]$. Then*

$$\lim_{t \to \infty} P[X(t) = j] = \frac{1}{\mu} \int_0^\infty q_j(t)dt = \frac{E[\text{time in } j \text{ in a cycle}]}{E(\text{length of cycle})}. \quad (3.36)$$

For a lattice F of span h and $0 \leq t < h$,

$$\lim_{n \to \infty} P[X(t + nh) = j] = (h/\mu) \sum_{n=0}^\infty q_j(t + nh) \quad (3.37)$$

$$= \frac{E[\# \text{ of } t + nh \ (n = 0, 1, 2, \ldots) \text{ in a cycle when state is } j]}{E[(\text{length of cycle})/h]}.$$

Proof: The function $g(t) = P(T_1 > t)$ for $t \geq 0$ is nonnegative and decreasing with $\int_0^\infty g(t) = E(T_1) = \mu < \infty$, hence, it is dRi. By the right continuous paths and the isolated states of $X(t)$, the function $q_j(t)$ is right continuous in t. By the standard analysis theory, $q_j(t)$ is Riemann integrable. Because $q_j(t) \leq g(t)$, $q_j(t)$ is dRi and, if F is lattice of span h, $\sum_{n=0}^\infty q_j(t + nh) \leq \sum_{n=0}^\infty g(nh) = E(T_1/h) < \infty$ for any $t \geq 0$. Thus, the conditions for (3.29) and (3.30) in the Key Renewal Theorem are verified with $z(t) = q_j(t)$.

Event $[X(t) = j]$ is T_1-shift invariant (§3.5), so $H(t) = P[X(t) = j]$ satisfies the renewal equation $H = q_j + H * F$, and hence $H = q_j + q_j * m$. By (3.29), for a nonlattice F, $\lim_{t\to\infty} P[X(t) = j] = (1/\mu) \int_0^\infty q_j(t)dt$, which implies (3.36). For a lattice F, (3.37) follows from (3.30). ◇

Long-run time average: Without assuming a nonlattice renewal cycle time distribution,

$$\lim_{t\to\infty} \frac{\text{Time in } j \text{ by time } t}{t} = \frac{E(\text{time in } j \text{ in a cycle})}{E(\text{length of cycle})}. \qquad (3.38)$$

Therefore, the long-run fraction of time when $X(t)$ is in j is the same as the limiting probability $\lim_{t\to\infty} P[X(t) = j]$ when F is nonlattice. The equality (3.38) is a much simpler result than (3.36) and follows from the limiting property of a cumulative renewal reward process with a reward earned at unit rate whenever the process is in state j.

Smith's theorem on a general state space: As in Theorem 3.26 but with $X(t)$ taking values in a general metric space S, for any bounded continuous function f on S, $q(t) = E[f(X(t)); T_1 > t]$ is dRi and

$$\lim_{t\to\infty} E[f(X(t))] = \frac{1}{\mu} \int_0^\infty q(t)dt = \frac{1}{\mu} E\left[\int_0^{T_1} f(X(t))dt\right] \qquad (3.39)$$

if F is nonlattice. For a lattice F of span h and $0 \le t < h$,

$$\lim_{n\to\infty} E[f(X(t + nh))] = \frac{h}{\mu} \sum_{n=0}^\infty q(t + nh) \quad (< \infty). \qquad (3.40)$$

Moreover, without the nonlattice assumption on F,

$$\lim_{t\to\infty} \frac{1}{t} \int_0^t f(X(s))ds = \frac{1}{\mu} E\left[\int_0^{T_1} f(X(t))dt\right]. \qquad (3.41)$$

The proof of (3.39) and (3.40) follows that of Theorem 3.26 using the T_1-shift invariance of process $f(X(t))$, and (3.41) is proved by an accumulative renewal reward argument with reward $\int_0^t f(X(s))ds$.

Stationary processes: A process $X(t)$ is called a stationary process if for any constant $T > 0$, the time-shifted process $X^T(t) = X(T + t)$ has the same distribution as the process $X(t)$. Note the distinction between a stationary process and a process with stationary increments. A stationary renewal process is not a stationary process.

Theorem 3.27 *Let $X(t)$ be a regenerative process. If the associated renewal cycle times are nonlattice and have a finite mean, then $X(t)$ becomes a stationary process as $t \to \infty$ in the sense that the time-shifted process $X^u(t) = X(u+t)$ converges to a stationary process in distribution as time shift $u \to \infty$.*

Proof: We need to show that all finite dimensional distributions of $X^u(t)$ converge to those of a stationary process. For $t_1 < t_2 < \cdots < t_n$, let
$$\bar{X}(u) = (X(u+t_1), X(u+t_2), \ldots, X(u+t_n)).$$
This is not a regenerative process, but is T_1-shift invariant. Although the Smith Theorem (Theorem 3.26) is stated for regenerative processes, its proof is valid for any T_1-shift invariant process, so it may be applied to $\bar{X}(u)$ to show that the finite dimensional distributions of $X^u(t)$ converge as $u \to \infty$. If the limits are the finite dimensional distributions of a process, then it must be stationary. The existence of this process is guaranteed by Kolmogorov's theorem, see §1.9. ◇

Steady state of a queuing system: The total queue length process $\bar{Q}(t)$ together with some other processes describe the evolution of a queuing system. For a stable queuing system, these processes are regenerative processes based on the iid cycle of the system. By Theorem 3.27, for a nonlattice stable queuing system, these processes will become stationary processes in the long run and their distributions become the limiting distributions. Then we will say the system is in the steady state.

3.8 Queue length distribution and PASTA

Total queue length distribution: For a stable queuing system, let π_j be the long-run fraction of time when there are j customers in the system. By (3.38),
$$\pi_j = \frac{E[\text{amount of time when } j \text{ in system during a cycle}]}{E[\text{length of a cycle}]}. \tag{3.42}$$

By the Smith Theorem, when the queuing system is nonlattice,
$$\pi_j = \lim_{t \to \infty} P[\bar{Q}(t) = j], \tag{3.43}$$

where $\bar{Q}(t)$ is the total queue length process. The numbers π_j form a probability distribution on the set of $j = 0, 1, 2, \ldots$, which may be called the steady-state total queue length distribution.

For a stable single-server system with arrival rate a and service rate b, and traffic intensity $\rho = a/b < 1$, as mentioned in §3.3 (after (3.9)), π_0 as the long-run fraction of idle time is given by

$$\pi_0 = 1 - \rho. \tag{3.44}$$

Expected queue length: Recall

$$Q = \lim_{t\to\infty} \frac{1}{t} \int_0^t Q(s)ds \quad \text{and} \quad \bar{Q} = \lim_{t\to\infty} \frac{1}{t} \int_0^t \bar{Q}(s)ds$$

are, respectively, the long-run time averages of the queue length and the total queue length defined in Proposition 3.13. It is easy to show from (3.42) and (3.10) that for a stable queuing system,

$$\bar{Q} = \sum_{j=0}^{\infty} j\pi_j, \tag{3.45}$$

and when there are k servers,

$$Q = \sum_{j=k}^{\infty} (j - k)\pi_j. \tag{3.46}$$

Thus, Q and $\bar{Q}$ are also equal to, respectively, the expected queue length and the expected total queue length in steady state.

Total queue length seen by arrivals: Let $A(j, n)$ be the number of arriving customers among the first n arrivals who see j in the system (not including themselves). When determining this number and a similar number for departures later, one assumes that all arrivals are numbered according to the order in which they occur, and so are all departures. The arrivals and departures occurring at the same time are numbered by different but consecutive integers, and arrivals are regarded as occurring before departures when they occur at the same time. Thus, the number of customers in the system seen by the mth arrival at time t includes all customers departing at t and those arriving at t with order $< m$, but does not include those arriving at t with order $\geq m$.

Let σ_n be the number of customers served in the nth renewal cycle of the queuing system. By considering a renewal reward process based on iid cycles σ_n and with one unit of reward earned whenever an arrival sees j in the system, we obtain the existence of the limit below:

$$\pi_j^A = \lim_{n\to\infty} \frac{A(j,n)}{n} = \frac{E[A(j,\sigma)]}{E(\sigma)}. \qquad (3.47)$$

Total queue length seen by departures: Let $D(j,n)$ be the number of departing customers among the first n departures who see j in the system (not including themselves). Note that by the description of $A(j,n)$, the number of customers in the system seen by the m-th departure at time t include all arrivals at t and all departures at t with order $> m$, but does not include the departures at t with order $\leq m$.

By considering one unit of reward generated whenever a departure sees j in the system, based on iid cycles σ_n, we see that the limits below exist:

$$\pi_j^D = \lim_{n\to\infty} \frac{D(j,n)}{n} = \frac{E[D(j,\sigma)]}{E(\sigma)}. \qquad (3.48)$$

Theorem 3.28 *(Limiting probabilities) Let $\bar{Q}_n^A$ and $\bar{Q}_n^D$ be, respectively, the total queue lengths seen by the nth arrival and the nth departure in a stable queuing system. Assume $P(X > Y) > 0$, where X and Y are, respectively, generic inter-arrival and service times. Then*

$$\forall j \geq 0, \quad \lim_{n\to\infty} P[\bar{Q}_n^A = j] = \pi_j^A \quad \text{and} \quad \lim_{n\to\infty} P[\bar{Q}_n^D = j] = \pi_j^D. \quad (3.49)$$

Proof: Let $X(t) = \bar{Q}_{[t]+1}^A$, where $[t]$ is the integer part of t. Then $X(t)$ is a regenerative process associated to the renewal cycles σ_n. Because $P(X > Y) > 0$, $P(\sigma = 1) > 0$ and hence σ has a lattice distribution of span $h = 1$. By (3.37) in the Smith Theorem, $P[\bar{Q}_n^A = j] = P[X(n) = j] \to E[A(j,\sigma)]/E(\sigma) = \pi_j^A$ as $n \to \infty$. This proves the equality for π_j^A. The equality for π_j^D can be proved in the same way. $\diamondsuit$

Theorem 3.29 *In a stable queuing system, the limiting total queue length distribution seen by arrivals and that seen by departures are the same, that is, $\pi_j^A = \pi_j^D$ for all $j \geq 0$.*

Proof: This follows directly from the fact that for a stable queuing system, any change of queue length from j to $j + 1$, when an arrival sees j in the system, must be matched by a change from $j + 1$ to j, when a departure sees j in the system. $\diamondsuit$

PASTA: A stable queuing system with Poisson arrivals is nonlattice because its interarrival distribution is so. By the theorem below, the steady-state total queue length distribution $\{\pi_j\}$ is the same as the limiting total queue length distribution $\{\pi_j^A\}$ seen by arrivals. This phenomenon is called PASTA (Poisson arrivals see time average).

Theorem 3.30 *In a stable queuing system with Poisson arrivals,*

$$\pi_j = \pi_j^A \text{ for all } j \geq 0.$$

Proof: Let $N(t)$ be the Poisson process of arriving customers at rate a and let $R(t)$ be the number of arrivals seeing j by time t. Then

$$\pi_j^A = \lim_{t\to\infty} \frac{R(t)}{N(t)} = \lim_{t\to\infty} \frac{R(t)/t}{N(t)/t} = \lim_{t\to\infty} \frac{R(t)/t}{a},$$

and hence $\lim_{t\to\infty} R(t)/t = a\pi_j^A$. This limit in fact holds in expectation, that is,

$$\lim_{t\to\infty} E[R(t)]/t = a\pi_j^A.$$

This follows from the lemma below, the Elementary Renewal Theorem (Theorem 3.4), and the fact that $R(t) \leq N(t)$.

Lemma 3.31 *Let x_n and y_n be two sequences of random variables such that $0 \leq x_n \leq y_n$, and as $n \to \infty$, $x_n \to \alpha$ and $y_n \to \beta$ a.s. for constants α and β. If $E(y_n) \to \beta$, then $E(x_n) \to \alpha$.*

To prove the lemma, note by Fatou's Lemma, which says that $\liminf_n E(z_n) \geq E(\liminf_n z_n)$ for any sequence of random variables $z_n \geq 0$, $\liminf_n E(x_n) \geq \alpha$ and $\liminf_n E(y_n - x_n) \geq (\beta - \alpha)$. On the other hand,

$$\beta = \lim_n E(y_n) \geq \liminf_n E(x_n) + \liminf_n E(y_n - x_n) \geq \alpha + (\beta - \alpha).$$

This implies $\liminf_n E(x_n) = \alpha$ and $\liminf_n E(y_n - x_n) = \beta - \alpha$. The latter implies $\lim_n E(y_n) - \limsup_n E(x_n) = \beta - \alpha$. Then $\limsup_n E(x_n) = \alpha$. This together with $\liminf_n E(x_n) = \alpha$ proves $\lim_n E(x_n) = \alpha$. The lemma is proved.

We now have, with S_n being the arrival time of the nth customer,

$$\pi_j^A = \lim_{t\to\infty} \frac{E[R(t)]}{at} = \lim_{t\to\infty} \frac{1}{at} E\left[\sum_{n=1}^{N(t)} 1_{\{j\}}(\bar{Q}(S_n-))\right]$$

$$= \frac{1}{a} E[\sum_{n=1}^{N(1)} 1_{\{j\}}(\bar{Q}(S_n-))] \quad \text{(may assume system in steady state)}$$

$$= \lim_{m\to\infty} \frac{1}{a} E\{\sum_{k=0}^{m-1} 1_{\{j\}}(\bar{Q}(\frac{k}{m}-))[N(\frac{k+1}{m}) - N(\frac{k}{m})]\} \quad ([0,1] \text{ is}$$

divided into m subintervals and left continuity of $\bar{Q}(t-)$ used here)

$$= \lim_{m\to\infty} \frac{1}{a} \sum_{k=0}^{m-1} E[1_{\{j\}}(\bar{Q}(\frac{k}{m}-))](\frac{a}{m})$$

$(N(\frac{k+1}{m}) - N(\frac{k}{m}))$ is independent of $\bar{Q}(\frac{k}{m}-)$ with mean $\frac{a}{m}$)

$$= \int_0^1 E[1_{\{j\}}(\bar{Q}(t-))]dt = \int_0^1 E[1_{\{j\}}(\bar{Q}(t))]dt = E[1_{\{j\}}(\bar{Q}(0))] = \pi_j$$

(steady state). $\diamondsuit$

Total queue length of a stable M/G/1: In a stable M/G/1 system, the numbers A_n of arrivals during successive service periods are iid. Let $a_j = P(A = j)$ for $j = 0, 1, 2, \ldots$. Then the limiting total queue length distribution seen by departures, π_j^D, satisfy the following recursive relation.

$$\pi_i^D = \pi_0^D a_i + \sum_{j=1}^{i+1} \pi_j^D a_{i+1-j} \quad \text{for } i \geq 0, \quad (3.50)$$

that is,

$$\pi_0^D = \pi_0^D a_0 + \pi_1^D a_0,$$
$$\pi_1^D = \pi_0^D a_1 + \pi_1^D a_1 + \pi_2^D a_0,$$
$$\pi_2^D = \pi_0^D a_2 + \pi_1^D a_2 + \pi_2^D a_1 + \pi_3^D a_0, \quad \ldots.$$

To show this, note that for $j \geq 1$, if a departure sees j and $i+1-j$ arrive during the next service, the next departure will see i. This implies that $\pi_j^D a_{i+1-j}$ is the fraction of departures that see i when the previous departures see j. Sum over j to get the fraction of departures that see i when the previous departures see at least 1. This plus $\pi_0^D a_i$, which is the fraction of departures that see i when the previous departures see 0, is equal to π_i^D. This proves (3.50). See Example 4.26 for a more formal proof of (3.50).

By Theorems 3.29 and 3.30, $\pi_j = \pi_j^A = \pi_j^D$, and by (3.44), $\pi_0 =$

$1 - \rho$. Thus, (3.50) may be used to compute π_j for $j \geq 1$ recursively. With Y being a generic service time,

$$\begin{aligned} a_j &= P(A = j) = E[P(A = j \mid Y)] = E[e^{-aY}\frac{(aY)^j}{j!}] \\ &= \frac{a^j(-1)^j}{j!}L_Y^{(j)}(a) \quad \text{for } j \geq 0, \end{aligned} \quad (3.51)$$

where a is the arrival rate and $L_Y^{(j)}(t)$ is the jth order derivative of the Laplace transform $L_Y(t)$ of Y. Note that the traffic intensity is $\rho = E(aY)$, the mean number of arrivals in a service period.

PASTA in a more general form: Let $X(t)$ be a regenerative process based on the renewal cycles of a stable queuing system with a Poisson process $N(t)$ of arrivals. By (3.39), $X(t)$ converges in distribution to some random variable $X(\infty)$ as $t \to \infty$ in the sense that for any bounded continuous function $f(x)$, $\lim_{t\to\infty} E[f(X(t))] = E[f(X(\infty))]$. The distribution of $X(\infty)$ is the steady-state distribution of $X(t)$.

The following theorem is a more general version of PASTA, which implies Theorem 3.30 with $X(t) = 1_{\{j\}}(\bar{Q}(t))$. Let $X^n(t) = X(S_n^Q + t)$ and $N^n(t) = N(S_n^Q + t) - N(S_n^Q)$, where S_n^Q is the nth renewal time of the queuing system. We will assume a stronger regenerative property: For any $n \geq 1$, the pair of the shifted processes $(X^n(t), N^n(t))$ have the same distribution as the original process pair $(X(t), N(t))$, and are independent of $\{S_1^Q, S_2^Q, \ldots, S_n^Q\}$ and the original process pair up to time S_n^Q. Moreover, we will also assume for any $u < v$, $N(v) - N(u)$ is independent of the process $X(t)$ up to time u.

Theorem 3.32 *Under the above assumptions, $X(S_n-) \overset{d}{\to} X(\infty)$ (convergence in distribution) as $n \to \infty$, where S_n are the successive arrival times of $N(t)$. Moreover, if $X(t) \geq 0$, then*

$$\lim_{n\to\infty} \frac{X(S_1-) + \cdots + X(S_n-)}{n} = E[X(\infty)]. \quad (3.52)$$

Proof: First assume $X(t) \geq 0$. By a simple renewal reward argument based on iid cycles σ_n with a reward $X(S_k-)$ at time k,

$$\lim_{n\to\infty} \frac{X(S_1-) + \cdots + X(S_n-)}{n} = \frac{E[X(S_1-) + \cdots + X(S_\sigma-)]}{E(\sigma)}. \quad (3.53)$$

Now assume $X(t)$ is bounded. We may actually assume $X(t) \leq 1$. Then the proof of (3.52) is essentially the same as that of Theorem 3.30 with some simple changes, for example, $R(t)$ is now $\sum_{n=1}^{N(t)} X(S_n-)$. For an unbounded $X(t)$, the left-hand side of (3.53) is equal to

$$\lim_{b\to\infty} \frac{E[X(S_1-)\wedge b + \cdots + X(S_\sigma-)\wedge b]}{E(\sigma)}$$
$$= \lim_{b\to\infty} E[X(\infty)\wedge b] = E[X(\infty)].$$

This proves (3.52) for an unbounded $X(t)$.

Now allow $X(t)$ to take values in a general metric space. The process $Y(t) = X(S_{[t]+1}-)$ is a regenerative process based on renewal cycles σ_n. Because of Poisson arrivals, σ_n is lattice of span $h = 1$. By (3.40), for any bounded continuous function $f(x)$,

$$\lim_{n\to\infty} E[f(X(S_n-))] = \lim_{n\to\infty} E[f(Y(n))]$$
$$= \frac{E[f(X(S_1-)) + \cdots + f(X(S_\sigma-))]}{E(\sigma)}.$$

The last expression above is just the left-hand side of (3.52) when $X(t)$ is replaced by $f(X(t))$, so it is equal to $E[f(X(\infty))]$. This shows that $X(S_n-)) \overset{d}{\to} X(\infty)$. ◇

Waiting times: Let $W(t)$ and $\bar{W}(t)$ be, respectively, the waiting time and the total waiting time of a fictitious customer entering a stable queuing system at time t, called the virtue waiting time and the virtue total waiting time of the system at time t. Then both $W(t)$ and $\bar{W}(t)$ are regenerative processes based on the renewal cycles of the queuing system. If the system is nonlattice, then by (3.39),

$$W(t) \overset{d}{\to} W(\infty) \quad \text{and} \quad \bar{W}(t) \overset{d}{\to} \bar{W}(\infty) \tag{3.54}$$

as $t \to \infty$ for some random variables $W(\infty)$ and $\bar{W}(\infty)$, whose distributions are, respectively, the distributions of the virtue waiting time and the virtue total waiting time of the system in steady state.

Let W_n and $\bar{W}_n$ be, respectively, the waiting time and the total waiting time of the nth customer who arrives at time S_n. Note that $W_n = W(S_n-)$ and $\bar{W}_n = \bar{W}(S_n-)$. Recall the long-run averages of the waiting and total waiting times, W and $\bar{W}$, are defined in §3.4 as

$$W = \lim_{n\to\infty} \frac{W_1 + \cdots + W_n}{n} \quad \text{and} \quad \bar{W} = \lim_{n\to\infty} \frac{\bar{W}_1 + \cdots + \bar{W}_n}{n}.$$

By Theorem 3.32, with Poisson arrivals,

$$W_n \overset{d}{\to} W(\infty) \quad \text{and} \quad \bar{W}_n \overset{d}{\to} \bar{W}(\infty) \tag{3.55}$$

as $n \to \infty$. Moreover,

$$W = E[W(\infty)] \quad \text{and} \quad \bar{W} = E[\bar{W}(\infty)]. \tag{3.56}$$

Note on non-Poisson arrivals: With non-Poisson arrivals, the state of a queuing system observed at arrival times of customers may have a distribution different from that observed at an arbitrary time in steady state. Consider the following rather trivial example. Assume the inter-arrival time is uniform on $(1, 2)$ and the service time is 1 unit. Then the total queue length $\bar{Q}_n$ observed by the nth arrival and his waiting time W_n are both zero, but $\bar{Q}(\infty)$ and $W(\infty)$, which are, respectively, the limits in distribution of the total queue length $\bar{Q}(t)$ and the virtue waiting time $W(t)$ as $t \to \infty$, are not almost surely equal to 0.

Exercise 3.24 In an M/G/1 system with exponential arrival rate 1 and uniform service time on $(0, 1)$, find the long-run fraction of time when there are at least 3 in system.

Exercise 3.25 Suppose customers arrive at a singer-server facility according to a Poisson process. The service time is uniform on $(0, 1)$. Suppose 20% customers find the server free. Determine the average waiting time (before service).

Exercise 3.26 Let $\bar{q}$ be the total queue length observed by a departure in a stable M/G/1 in steady state. By Theorems 3.28, 3.29, and 3.30, $\bar{q}$ has the steady-state total queue length distribution, and so by (3.45), $\bar{Q} = E(\bar{q})$. Compute $E(\bar{q})$ to re-derive the formula for $\bar{Q}$ in Theorem 3.17.
Hint: Let $\bar{q}'$ be the total queue length observed by the next departure and let A be the number of arrivals during his service. Then

$$\bar{q}' = \bar{q} - \delta + A,$$

where $\delta = 1$ if $\bar{q} > 0$ and $\delta = 0$ if $\bar{q} = 0$. Taking the expectation of the above will not yield any useful information about $E(\bar{q})$ because $E(\bar{q}') = E(\bar{q})$. However, you may first square the above equation and then take expectation.

Chapter 4

Discrete time Markov chains

4.1 Markov property and transition probabilities

Markov property: A discrete time Markov chain X_n is a process of discrete time and discrete states which satisfies the following Markov property:

$$P(X_{n+1} = j \mid X_0 = i_0, X_1 = i_1, \ldots, X_n = i_n)$$
$$= \quad P(X_{n+1} = j \mid X_n = i_n) \qquad (4.1)$$

for any integer time $n \geq 0$ and states $j, i_0, i_1, \ldots, i_n$, provided

$$P(X_0 = i_0, \ldots, X_n = i_n) > 0.$$

It means that given the present (time n), the future distribution of the process (at time $n + 1$) will not depend on the past (at times $0, 1, 2, \ldots, n - 1$).

Transition probability and time homogeneity: The conditional probability

$$p_{ij} \quad = \quad P(X_{n+1} = j \mid X_n = i) \qquad (4.2)$$

is called the (1-step) transition probability from state i to state j at time n. In many applications, the quantity is independent of the time n, and then the Markov chain is called time homogeneous. In the sequel, a MC (Markov chain) is always assumed to be time homogeneous unless explicitly stated otherwise.

The transition probabilities p_{ij} satisfy:

$$0 \leq p_{ij} \leq 1 \text{ for all } i \text{ and } j, \quad \text{and} \quad \sum_j p_{ij} = 1 \text{ for all } i. \qquad (4.3)$$

Transition probability matrix: The matrix formed by the transition probabilities, $\mathbf{P} = \{p_{ij}\}$ (possibly of infinite size), is called the

(1-step) transition probability matrix. A transition probability matrix is characterized by the property that all matrix elements are nonnegative and all rows sum to 1.

Distribution of MC: The distribution of the MC X_n as a process is completely determined by the transition probabilities p_{ij} and the initial distribution $p_i = P(X_0 = i)$ as

$$P(X_0 = i_0, X_1 = i_1, \ldots, X_n = i_n) = p_{i_0} p_{i_0 i_1} \cdots p_{i_{n-1} i_n}. \qquad (4.4)$$

This may be proved using the Markov property as follows:

$$
\begin{aligned}
&P(X_0 = i_0, X_1 = i_1, \ldots, X_n = i_n) \\
=\ &P(X_n = i_n \mid X_0 = i_0, X_1 = i_1, \ldots, X_{n-1} = i_{n-1}) \\
&\times P(X_0 = i_0, X_1 = i_1, \ldots, X_{n-1} = i_{n-1}) \\
=\ &P(X_n = i_n \mid X_{n-1} = i_{n-1}) \\
&\times P(X_0 = i_0, X_1 = i_1, \ldots, X_{n-1} = i_{n-1}) \quad \text{(by (4.1))} \\
=\ &p_{i_{n-1} i_n} P(X_0 = i_0, X_1 = i_1, \ldots, X_{n-1} = i_{n-1}) \quad \text{(by (4.2))} \\
=\ &p_{i_{n-1} i_n} p_{i_{n-2} i_{n-1}} P(X_0 = i_0, X_1 = i_1, \ldots, X_{n-2} = i_{n-2}) \\
=\ &\cdots = p_{i_{n-1} i_n} p_{i_{n-2} i_{n-1}} \cdots p_{i_0 i_1} p_{i_0}.
\end{aligned}
$$

Existence of MC: An MC X_n may be constructed from a given set of transition probabilities p_{ij} and initial distribution p_i, that is, a set of numbers $p_{ij} \geq 0$ satisfying (4.3) and $p_i \geq 0$ satisfying $\sum_i p_i = 1$, such that (4.4) holds. This follows from Kolmogorov's theorem for the existence of stochastic processes, see §1.9. Moreover, such an MC is unique in distribution because its distribution is completely determined by p_{ij} and p_i in (4.4).

A more symmetric form of Markov property: For any integers $n > 0$ and $m > 0$, and states $i_0, i_1, \ldots, i_n, j_1, \ldots, j_m$, we have

$$
\begin{aligned}
&P(X_{n+1} = j_1, \ldots, X_{n+m} = j_m \mid X_0 = i_0, X_1 = i_1, \ldots, X_n = i_n) \\
=\ &P(X_{n+1} = j_1, \ldots, X_{n+m} = j_m \mid X_n = i_n) \qquad\qquad (4.5) \\
=\ &P(X_1 = j_1, \ldots, X_m = j_m \mid X_0 = i_n). \qquad\qquad\qquad (4.6)
\end{aligned}
$$

The first equality (4.5) is a consequence of the Markov property (4.1), which means that given the present (X_n), the future ($X_{n+1}, \ldots, X_{n+m}$) is independent of the past ($X_0, \ldots, X_{n-1}$), but now the future and the past are put in a more symmetric form, whereas the second equality (4.6) is a consequence of the time homogeneity.

We will prove this for $m = 2$. The proof of a general $m > 0$ is similar. First, we note that in the Markov property (4.1), the event $[X_k = i_k]$ for $0 \leq k \leq n - 1$ may be added to the condition on the right-hand side (see Exercise 4.1). Then

$$
\begin{aligned}
& P(X_{n+1} = j_1, X_{n+2} = j_2 \mid X_0 = i_0, X_1 = i_1, \ldots, X_n = i_n) \\
=\ & P(X_{n+2} = j_2 \mid X_0 = i_0, X_1 = i_1, \ldots, X_n = i_n, X_{n+1} = j_1) \\
& \times P(X_{n+1} = j_1 \mid X_0 = i_0, X_1 = i_1, \ldots, X_n = i_n) \qquad (4.7) \\
=\ & P(X_{n+2} = j_2 \mid X_n = i_n, X_{n+1} = j_1) P(X_{n+1} = j_1 \mid X_n = i_n) \\
=\ & P(X_{n+1} = j_1, X_{n+2} = j_2 \mid X_n = i_n).
\end{aligned}
$$

This shows that (4.5) with $m = 2$ follows from (4.1). To derive (4.6) with $m = 2$ from the time homogeneity, note that (4.7) above may be written as

$$
\begin{aligned}
p_{j_1 j_2} p_{i_n j_1} &= P(X_2 = j_2 \mid X_0 = i_n, X_1 = j_1) P(X_1 = j_1 \mid X_0 = i_n) \\
&= P(X_1 = j_1, X_2 = j_2 \mid X_0 = i_n).
\end{aligned}
$$

A more useful form of Markov property: Any event A_n determined by the process in the past (before time n) can be written as a union of events of the form $[X_0 = i_0, X_1 = i_1, \ldots, X_{n-1} = i_{n-1}]$ and any event B_n determined by the process in the future (after time n) may be written as a union of events of the form $[X_{n+1} = j_1, \ldots, X_{n+m} = j_m]$. Let B_0 be the event obtained from B_n by replacing each X_{n+i} by X_i, called the event B_n time shifted backward by n. For example, if $B_n = [X_{n+1} = j_1, \ldots, X_{n+m} = j_m]$, then $B_0 = [X_1 = j_1, \ldots, X_m = j_m]$. The Markov property (4.5) and the time homogeneity (4.6) can now be written in the following more useful form:

$$
P(B_n \mid A_n, X_n = i_n) = P(B_n \mid X_n = i_n) = P(B_0 \mid X_0 = i_n). \quad (4.8)
$$

The proof of (4.8) is left as an exercise (see Exercise 4.1). Note that in the Markov property (4.8), A_n may be an event determined by the MC before and at time n, which is a union of events of the form $[X_0 = i_0, X_1 = i_1, \ldots, X_n = i_n]$ with i_n fixed as in (4.8). Similarly, B_n may be an event determined by the MC at and after time n.

Markov property in expectation form: Let Y_n be a random variable determined by the MC before time n and let Z_n be an event

determined by the MC after time n. Then $Y_n = f(X_0, X_1, \ldots, X_{n-1})$
and $Z_n = g(X_{n+1}, X_{n+2}, \ldots, X_{n+m})$ for some Borel functions f and g.
We have

$$E[Z_n \mid Y_n, X_n] = E[Z_n \mid X_n] = E[Z_0 \mid X_0], \qquad (4.9)$$

where $Z_0 = g(X_1, X_2, \ldots, X_m)$ is the event Z_n time shifted backward
by n. This expectation form of the Markov property follows from (4.8)
and the definition of the conditional expectation given in §1.11.

Conditional independence of past and future given present:
The Markov property without time homogeneity, the first equality in
(4.8), may also be written as

$$P(A_n \cap B_n \mid X_n = i) = P(A_n \mid X_n = i)P(B_n \mid X_n = i). \qquad (4.10)$$

This means that given the present $[X_n = i]$, the past A_n and future B_n
are independent. The proof of (4.10) is left to Exercise 4.2.

Strong Markov property: An MC in fact satisfies a stronger form
of Markov property, in which the constant time n in (4.8) is replaced
by a stopping time τ with $P(\tau < \infty) > 0$, that is,

$$\begin{aligned} P(B_\tau \mid A_\tau, X_\tau = i, \tau < \infty) &= P(B_\tau \mid X_\tau = i, \tau < \infty) \\ &= P(B_0 \mid X_0 = i) \qquad (4.11) \end{aligned}$$

for any events A_τ and B_τ determined by the process before time τ
and after time τ respectively, where B_0 is the event B_τ time shifted
backward by τ. This is called the strong Markov property. In particular,

$$p_{ij} = P(X_{\tau+1} = j \mid X_\tau = i, \tau < \infty).$$

Note that A_τ is a union of events of the form

$$[X_0 = i_0, \ldots, X_{n-1} = i_{n-1}, \tau = n], \qquad (4.12)$$

B_τ is a union of events of the form

$$[X_{\tau+1} = j_1, \ldots, X_{\tau+m} = j_m, \tau < \infty], \qquad (4.13)$$

and $B_0 = [X_1 = j_1, \ldots, X_m = j_m]$ if B_τ is given in (4.13).
The proof of the strong Markov property is left to Exercise 4.3.

Exercise 4.1 (a) Show that in the Markov property (4.1), one may add the event $[X_k = i_k]$ for $0 \le k \le n-1$ to the condition on the right-hand side, that is, show from (4.1) that for $0 \le k < n$,

$$P(X_{n+1} = j \mid X_n = i_n) = P(X_{n+1} = j \mid X_k = i_k, X_n = i_n).$$

In fact, more than one such event may be thus added.
(b) Prove (4.8) from the Markov property (4.1) and the time homogeneity.

Exercise 4.2 Prove the conditional independence (4.10).

Exercise 4.3 Prove the strong Markov property (4.11).
Hint: It suffices to prove (4.11) for A_τ and B_τ in (4.12) and (4.13).

4.2 Examples of discrete time Markov chains

Example 4.1 (A simple random walk) As in Exercise 1.2, but with a change of notation, let ξ_n be iid with $P(\xi = 1) = p$ and $P(\xi = -1) = q = 1 - p$ for $0 < p < 1$. Let X_0 be an integer-valued random variable independent of the sequence ξ_n, and let

$$X_n = X_0 + \xi_1 \cdots + \cdots + \xi_n$$

for $n \ge 1$. The process X_n is called a simple random walk on integers. Since $X_{n+1} = X_n + \xi_{n+1}$ with ξ_{n+1} independent of $X_0, X_1, \ldots, X_n$, it follows that given X_n, X_{n+1} is independent of $X_0, X_1, \ldots, X_{n-1}$, so X_n is a discrete time MC with transition probabilities

$$p_{i\,i+1} = p, \quad p_{i\,i-1} = q \quad \text{and all other } p_{ij} = 0.$$

These transition probabilities may be conveniently depicted in the following transition diagram:

$$\cdots \overset{q}{\leftarrow} (-1) \overset{p}{\underset{q}{\rightleftarrows}} (0) \overset{p}{\underset{q}{\rightleftarrows}} (1) \overset{p}{\underset{q}{\rightleftarrows}} (2) \overset{p}{\rightarrow} \cdots \overset{q}{\leftarrow} (n) \overset{p}{\rightarrow} \cdots.$$

When $p = 1/2$, the MC X_n is called a symmetric random walk.

Example 4.2 (Absorbing or reflecting boundary) Let $N > 0$ be an integer. We may restrict a simple random walk on integers between 0 and N, and impose the absorbing condition at boundary states 0 and N, so that when the process enters either state, it will remain there forever. The modified process X_n satisfies $X_{n+1} = X_n + \zeta_n$, where $\zeta_n = \xi_{n+1}$ if $0 < X_n < N$, $\zeta_n = 0$ if $X_n = 0$ or N. Thus, given X_n, X_{n+1} is independent of $X_0, X_1, \ldots, X_{n-1}$, and hence X_n is still an MC, called a simple random walk with absorbing boundary at 0 and N. Its transition probabilities p_{ij} are as in Example 4.1 except now $p_{00} = p_{NN} = 1$. Its transition diagram is as follows:

$$\overset{1}{\hookrightarrow} (0) \overset{q}{\leftarrow} (1) \overset{p}{\underset{q}{\rightleftarrows}} (2) \overset{p}{\rightarrow} \cdots \overset{q}{\leftarrow} (N-1) \overset{p}{\rightarrow} (N) \overset{1}{\hookleftarrow}.$$

We may also impose the reflecting boundary condition so that when the process enters a boundary state 0 or 1, it will next move inside $[0, N]$. Its transition probabilities are as in Example 4.1 except now $p_{01} = p_{N\,N-1} = 1$. Its transition diagram is depicted below:

$$(0) \overset{1}{\underset{q}{\rightleftarrows}} (1) \overset{p}{\underset{q}{\rightleftarrows}} (2) \overset{p}{\rightarrow} \cdots \overset{q}{\leftarrow} (N-1) \overset{p}{\underset{1}{\rightleftarrows}} (N).$$

Example 4.3 (A simple branching process) In a population, each individual lives 1 unit of time and then produces i offsprings with probability p_i for $i = 0, 1, 2, \ldots$, independently of other individuals. Let X_n be the size of the population at time n, called a simple branching process with offspring distribution $p = (p_0, p_1, p_2, \ldots)$. Let ξ_k be the number of offsprings produced by the kth individual in nth generation. Then ξ_k are iid with distribution p and

$$X_{n+1} = \sum_{k=1}^{X_n} \xi_k.$$

Because the above is clearly independent of $X_0, X_1, \ldots, X_{n-1}$ given X_n, so X_n is an MC with transition probabilities given by $p_{00} = 1$, $p_{0j} = 0$ for $j > 0$, $p_{1j} = p_j$, $p_{2j} = \sum_{k=0}^{j} p_k p_{j-k}$ and

$$p_{ij} = \sum_{m_1 + m_2 + \cdots + m_i = j} p_{m_1} p_{m_2} \cdots p_{m_i}$$

for $i > 1$ and $j \geq 0$. Note that for fixed $i > 1$, the distribution $(p_{i0}, p_{i1}, p_{i2}, \ldots)$ is the i-fold convolution of the offspring distribution p regarded as a pmf.

Example 4.4 (An inventory model) Let $0 \leq s < S$ be two integers. The inventory level of a certain commodity is examined and replenished periodically. If the level at the end of a period is $\leq s$, then it is replenished immediately to the level S, otherwise, no replenishment takes place. This is called an (s, S) inventory policy.

Let X_n be the inventory level at the end of nth period (before replenishment). Let D_n be the quantities of the demand in the nth period. Assume D_n are iid with distribution $a_i = P(D = i)$ for $i = 0, 1, 2, \ldots$. Assume also that the part of demand exceeding the inventory level is lost. Then

$$X_{n+1} = \begin{cases} (X_n - D_{n+1})_+ & \text{if } X_n > s \\ (S - D_{n+1})_+ & \text{if } X_n \leq s \end{cases} \quad (x_+ = \max(x, 0)),$$

and hence X_n is an MC on states $0, 1, 2, \ldots, s, \ldots, S$. For $i \geq 0$, let $b_i = P(D \geq i) = \sum_{k=i}^{\infty} a_i$, and set $a_i = 0$ for $i < 0$. Then the transition probabilities are given by

$$p_{ij} = \begin{cases} a_{S-j} & \text{if } i \leq s \text{ and } j > 0 \\ b_S & \text{if } i \leq s \text{ and } j = 0 \\ a_{i-j} & \text{if } s < i \leq S \text{ and } j > 0 \\ b_i & \text{if } s < i \leq S \text{ and } j = 0. \end{cases}$$

Example 4.5 (A discrete queuing model) Customers arrive at a single-server station and are served in the order of arrivals. The service starts at each integer time when there is a customer waiting, and lasts 1 unit time. Let A_n be the number of arrivals in time interval $(n-1, n]$ for $n = 1, 2, 3, \ldots$. Assume A_n are iid with distribution $a_i = P(A = i)$ for $i = 0, 1, 2, \ldots$. Let X_n be the number of customers in the station at time n (not including the customer whose service has just been completed). Then

$$X_{n+1} = \begin{cases} X_n + A_{n+1} - 1 & \text{if } X_n > 0 \\ A_{n+1} & \text{if } X_n = 0. \end{cases}$$

Thus, X_n is an MC on states $0, 1, 2, \ldots$ with transition probability matrix

$$P = \begin{bmatrix} a_0 & a_1 & a_2 & a_3 & a_4 & \cdot \\ a_0 & a_1 & a_2 & a_3 & a_4 & \cdot \\ 0 & a_0 & a_1 & a_2 & a_3 & \cdot \\ 0 & 0 & a_0 & a_1 & a_2 & \cdot \\ 0 & 0 & 0 & a_0 & a_1 & \cdot \\ \cdot & \cdot & \cdot & \cdot & \cdot & \cdot \end{bmatrix},$$

noting the first row and the first column are associated to state 0.

Alternatively, we may let X_n be the total queue length of an $M/G/1$ system (defined in §3.4) at the end of the nth service. In this case, the numbers A_n of arrivals in successive service periods are iid with distribution $a_i = P(A = i)$ given by (3.51) in §3.8, and X_n is an MC with the same transition probabilities as above.

Exercise 4.4 Assume in a $(2,5)$-inventory policy, the demand in a period is uniform on $0, 1, 2, 3, 4$. Find the transition probabilities.

Exercise 4.5 Suppose customers arrive at a singer-server station according to a Poisson process of rate 1 per minute. The service time is uniform on $(0, 1)$. Assume the station has only two waiting spaces and a customer arriving at a full station will not enter. Let X_n be the number of customers in the station at the end of nth service (not including the customer whose service is just completed). Then X_n is an MC by the same reason as in Example 4.5. Determine its states and its transition probabilities.

Exercise 4.6 A salesman travels among three cities 1, 2, and 3. He spends one unit of the time in each city and then moves to another city. Suppose $1/3$ of the time he returns to the city he last visited and $2/3$ of the time he goes to a different city.
(a) Let $X_n = i$ if he is in city i at time n. Show X_n is not an MC.
(b) Now let $X_n = (i, j)$ if he is at city i at time $(n-1)$ and at city j at time n. Show that X_n is an MC, and determine transition probabilities.

4.3   Multi-step transition and reaching probabilities

Multi-step transition probabilities: For $m \geq 2$, the m-step transition probabilities are defined by

$$p_{ij}^{(m)} = P(X_m = j \mid X_0 = i) = P(X_{n+m} = j \mid X_n = i), \qquad (4.14)$$

where the second equality above holds for any integer $n > 0$ and is the consequence of the time homogeneity. The 1-step transition probabilities are of course $p_{ij}^{(1)} = p_{ij}$. It is convenient to set the 0-step transition probabilities $p_{ij}^{(0)}$ as Kronecker's delta δ_{ij} defined by

$$\delta_{ij} = \begin{cases} 1, & \text{if } i = j \\ 0, & \text{otherwise.} \end{cases}$$

By the total probability law applied to the conditional probability $P(\cdot \mid X_0 = i)$, it is easy to show the following recursive relation: for $m > 1$,

$$p_{ij}^{(m)} = \sum_k p_{ik} p_{kj}^{(m-1)}. \tag{4.15}$$

In particular, the 2-step transition probabilities are $p_{ij}^{(2)} = \sum_k p_{ik} p_{kj}$.

Chapman-Kolmogorov identity: $p_{ij}^{(m+n)} = \sum_k p_{ik}^{(m)} p_{kj}^{(n)}$.
    This follows easily from (4.15). Roughly speaking, it means that for the MC to go from i to j in $m + n$ steps, it has to go to an arbitrary state k in m steps and then from k to j in n steps.

Multi-step transition probability matrix: $\mathbf{P}^{(m)} = \{p_{ij}^{(m)}\}$ satisfies $\mathbf{P}^{(m+n)} = \mathbf{P}^{(m)}\mathbf{P}^{(n)}$. Thus, $\mathbf{P}^{(m)} = \mathbf{P}^m$ (m-fold matrix product).

Distribution of MC at time n: Using the n-step transition probabilities $\{p_{ij}^{(n)}\}$ and the initial distribution $p_i = P(X_0 = i)$, the distribution of MC at time n is given by

$$P(X_n = j) = \sum_i p_i p_{ij}^{(n)}. \tag{4.16}$$

In matrix form, this is $\mathbf{p}_n = \mathbf{p}_0 \mathbf{P}^n$, where $\mathbf{p}_n$ is the row vector of distribution at time n.

Hitting time: For a state i of an MC X_n, let

$$\tau_i = \inf\{n > 0; \ X_n = i\}. \tag{4.17}$$

This is the first time $n > 0$ when the MC visits state i, called the hitting time of state i, and is a stopping time of the MC. It is defined to be ∞ if $X_n \neq i$ for all $n > 0$.

Reaching probabilities: For any state i of an MC X_n, let P_i be the conditional probability $P(\cdot \mid X_0 = i)$. For $n > 0$, let

$$\begin{aligned} f_{ij}^{(n)} &= P_i(\tau_j = n) \\ &= P(X_1 \neq j, \, X_2 \neq j, \, \ldots, \, X_{n-1} \neq j \, X_n = j \mid X_0 = i). \end{aligned} \tag{4.18}$$

This is the probability of first reaching j from i at time n and can be computed recursively from

$$f_{ij}^{(1)} = p_{ij} \quad \text{and} \quad f_{ij}^{(n)} = \sum_{k \neq j} p_{ik} f_{kj}^{(n-1)} \quad \text{for } n > 1. \tag{4.19}$$

The above may be written in the following matrix form, which is more convenient in actual computation: For fixed j, let $p_{\cdot j}$ and $f_{\cdot j}^{(n)}$ be, respectively, the jth columns of the matrices $\mathbf{P} = \{p_{ij}\}$ and $\{f_{ij}^{(n)}\}$, and let $\mathbf{P}_{(j)}$ be the matrix $\mathbf{P}$ with column j set to zero. Then for $n > 1$,

$$f_{\cdot j}^{(n)} = [\mathbf{P}_{(j)}]^{n-1} p_{\cdot j} \tag{4.20}$$

because $f_{\cdot j}^{(n)} = [\mathbf{P}_{(j)}] f_{\cdot j}^{(n-1)} = [\mathbf{P}_{(j)}]^2 f_{\cdot j}^{(n-2)} = \cdots = [\mathbf{P}_{(j)}]^{n-1} p_{\cdot j}$.
   Let

$$f_{ij} = P_i(\tau_j < \infty) = P_i(X_n = j \text{ for some } n \geq 1). \tag{4.21}$$

This is the probability of reaching j from i and is given by

$$f_{ij} = \sum_{n=1}^{\infty} f_{ij}^{(n)}. \tag{4.22}$$

Example 4.6   Consider the MC on states $0, 1, 2, 3$ with transition probability matrix

$$\mathbf{P} = \begin{bmatrix} 0.1 & 0.2 & 0.3 & 0.4 \\ 0.2 & 0.2 & 0.3 & 0.3 \\ 0.2 & 0.3 & 0.3 & 0.2 \\ 0.5 & 0.2 & 0.1 & 0.2 \end{bmatrix}.$$

Note that the first row and column are associated with state 0. Find the probability that the MC starting at state 0 reaches state 3 at time 5 but not earlier.

Solution: The 4th column of $\mathbf{P}$ are transition probabilities to state 3.

$$f_{\cdot 3}^{(5)} = [\mathbf{P}_{(4)}]^{5-1} p_{\cdot 3} = \begin{bmatrix} 0.1 & 0.2 & 0.3 & 0 \\ 0.2 & 0.2 & 0.3 & 0 \\ 0.2 & 0.3 & 0.3 & 0 \\ 0.5 & 0.2 & 0.1 & 0 \end{bmatrix}^4 \begin{pmatrix} 0.4 \\ 0.3 \\ 0.2 \\ 0.2 \end{pmatrix} = \begin{pmatrix} 0.0639 \\ 0.0728 \\ 0.0829 \\ 0.0764 \end{pmatrix}$$

(using MATLAB®). The desired probability is $f_{03}^{(5)} = 0.0639$.

Exercise 4.7 Let X_n be an MC on states $0, 1, 2$ with the transition matrix and initial distribution given below

$$\mathbf{P} = \begin{bmatrix} .3 & .3 & .4 \\ .2 & .7 & .1 \\ .2 & .3 & .5 \end{bmatrix} \quad \text{and} \quad P(X_0 = \begin{cases} 0 \\ 1 \\ 2 \end{cases}) = \begin{cases} 0.2 \\ 0.2 \\ 0.6 \end{cases}$$

Find $P(X_4 = 1)$.

Exercise 4.8 For a simple random walk X_n with absorbing boundaries at 0 and 4, and step-up probability $p = 2/3$, assume that initially it is uniformly distributed on states $1, 2, 3$.
(a) Find the probability that $X_5 = 0$.
(b) Find the probability of the process reaching 0 at time 5 but not earlier.

Exercise 4.9 A barber shop has three chairs and one barber. Customers arrive according to a Poisson process of rate 2 and they will not enter if all three chairs are taken. Suppose each haircut takes one unit of time. Let X_n be the number of customers in the barber shop at time n (not including the one whose haircut is just completed). Find:
(a) the probability that the barber is free at time 4 given 2 customers at time 0; and
(b) the probability that the barber first becomes free at time 4 given 2 customers at time 0.

4.4 Classes, recurrence, and transience

Accessibility and communication: Let X_n be a MC. Recall P_i is the conditional probability given $X_0 = i$ and $f_{ij} = P_i(X_n = j$ for some $n > 0)$. A state j is said to be accessible from state i, denoted $i \to j$, if $f_{ij} > 0$, that is, if there is a positive probability of reaching j from i. It is easy to see that the following statements are equivalent:

(a) $i \to j$;
(b) $p_{ij}^{(n)} > 0$ for some $n \geq 0$; and
(c) $P_i(\tau_j < \infty) > 0$, where τ_j is the hitting time of j defined by (4.17).

Two states i and j are said to communicate, written $i \leftrightarrow j$, if $i \rightarrow j$ and $j \rightarrow i$.

Classes: The state space S of an MC X_n is the set of all states. It is divided into several subsets $C_1, C_2, \ldots, C_i, \ldots$, called classes, such that any two states within the same class communicate, but states from different classes do not.

A state i is called absorbing if $p_{ii} = 1$. Then

$$P_i(X_n = i \text{ for all } n > 0) = 1,$$

that is, the MC starting in i will never leave i. It is clear that an absorbing state forms a class by itself. The random walk on $\{0, 1, , \ldots, N\}$ with absorbing boundary has two absorbing states: 0 and N, and has three classes: $\{0\}$, $\{N\}$, and $\{1, 2, \ldots, (N-1)\}$.

Note that it may be possible to access one class from another, but there is no two-way communication between classes.

Irreducibility: An MC is called irreducible if its state space has only one class, that is, if any two states communicate. Some examples of irreducible MCs are listed below.

(a) A simple random walk on all integers or on $\{0, 1, 2, \ldots, N\}$ with reflecting boundary.

(b) Inventory level in an (s, S)-policy (Example 4.4) with demand distribution satisfying $P(D = j) > 0$ for $1 \leq j \leq s + 1$.

(c) The MC in a discrete queuing model (Example 4.5) with arrival distribution satisfying $P(A = 0) > 0$ and $P(A = j) > 0$ for some $j > 1$.

Recurrence and transience: A state i is called recurrent if $f_{ii} = 1$, that is, if the MC X_n starting from i will return to i with probability 1. Otherwise, the state i is called transient. An MC is called recurrent or transient if all its states are so.

Let E_i denote the conditional expectation given $X_0 = i$, that is, $E_i(Y) = E(Y \mid X_0 = i)$ for any random variable Y, and let N_i be the number of times $n > 0$ when $X_n = i$.

Theorem 4.7 *If i is recurrent, then starting from i, the MC will return to i infinitely many times with probability 1, that is, $P_i(N_i = \infty) = 1$. If i is transient, then the number of returns to i has a finite mean and hence it is finite with probability 1, that is, $E_i(N_i) < \infty$ and $P_i(N_i < \infty) = 1$.*

Proof: If i is recurrent, then from each return to i, the MC will return again. This implies that the MC will return to i infinitely many times. Now consider running the MC starting from a transient state i as a trial and call it a success if no return. The success probability p is positive because i is transient. Then the number of runs until the first no return, as a geometric random variable, has a finite mean. The theorem is proved. This type of informal argument can be made precise by the strong Markov property at successive return times as demonstrated below.

First assume i is recurrent. Let $\tau = \tau_i$. Then for any integer $n > 0$,

$$P_i(\text{return at least } n \text{ times})$$
$$= P(\text{return once and then at least } n - 1 \mid X_0 = i)$$
$$= P(\text{return at least } n - 1 \text{ after } \tau \mid X_0 = i, \tau < \infty, X_\tau = i)$$
$$\times P(\tau < \infty \mid X_0 = i) \quad (X_\tau = i \text{ on } [\tau < \infty])$$
$$= P(\text{return at least } n - 1 \text{ after } \tau \mid X_0 = i, \tau < \infty, X_\tau = i)$$
$$(\text{by recurrence, } P_i(\tau < \infty) = 1)$$
$$= P(\text{return at least } n - 1 \text{ times} \mid X_0 = i)$$
$$(\text{by the strong Markov property (4.11)})$$
$$= P_i(\text{return at least } n - 1 \text{ times})$$
$$= \cdots = P_i(\text{return at least } 1 \text{ time}) = 1.$$

Letting $n \uparrow \infty$ shows that $P_i(\text{return infinitely many times}) = 1$.

Now assume i is transient. Let N be the number of times when state i is visited, and let $\tau^1 = \tau_i, \tau^2, \tau^3, \ldots$, be the times of successive visits to i. Then $P_i(N \geq 1) = P_i(\tau^1 < \infty) = f_{ii}$, and for $n > 1$,

$$P_i(N \geq n) = P(\tau^1 < \infty, \tau^2 < \infty, \ldots, \tau^n < \infty \mid X_0 = i)$$
$$= P_i(\tau^1 < \infty)P(\tau^2 < \infty, \ldots, \tau^n < \infty \mid X_0 = i, \tau^1 < \infty)$$
$$= f_{ii}P(\tau^2 < \infty, \ldots, \tau^n < \infty \mid X_0 = i, \tau^1 < \infty, X_{\tau^1} = i)$$
$$= f_{ii}P_i(\tau^1 < \infty, \ldots, \tau^{n-1} < \infty) \quad (\text{strong Markov property})$$
$$= f_{ii}^2 P_i(\tau^1 < \infty, \ldots, \tau^{n-2} < \infty) = \cdots = f_{ii}^n.$$

Thus, $P_i(N = n) = f_{ii}^n - f_{ii}^{n+1} = f_{ii}^n(1 - f_{ii})$. This shows that N is a geometric random variable minus 1 and hence has a finite mean. $\diamond$

Some useful facts about recurrence and transience: Later it will be shown that the states in the same class are either all recurrent or all transient, so it makes sense to talk about recurrent or transient classes.

A class is called closed if it is not possible to access another class from inside it, that is, $i \not\to j$ for any state i in the class and any j outside. It is easy to see that if a class is not closed, then it must be transient. On the other hand, a closed class with finitely many states is recurrent, because if not, then the MC would have nowhere to go after finitely many visits to all transient states. The same reason also shows that an MC with finitely many states must have a recurrent state. These informal arguments can be made precise by using the strong Markov property as in the proof of Theorem 4.7; see Exercise 4.13.

As an example, for the simple random walk on states $0, 1, \ldots, N$ with absorbing boundary, the class $\{1, \ldots, N - 1\}$ is not closed, so it must be transient. On the other hand, if the boundary is reflecting, then the random walk has a single closed class that is recurrent.

Example 4.8 Consider the MC on states $0, 1, 2, 3, 4, 5$ with transition probability matrix $\mathbf{P}$ given below. Identify its classes, and determine which states are recurrent and which are transient.

$$\mathbf{P} = \begin{bmatrix} 0.3 & 0 & 0 & 0.4 & 0.3 & 0 \\ 0.3 & 0 & 0.2 & 0.2 & 0.3 & 0 \\ 0 & 0 & 0 & 1 & 0 & 0 \\ 0 & 0 & 0.4 & 0.2 & 0.4 & 0 \\ 0 & 0 & 0 & 0 & 0.5 & 0.5 \\ 0 & 0 & 0 & 0 & 0.2 & 0.8 \end{bmatrix}.$$

Solution: Since the second column contains only zeros, state 1 cannot be accessed from any state, so it forms a class by itself, which is not closed because $1 \to 0$, so 1 is transient. State 0 can only be accessed from 1, so it too forms a class by itself, which is not closed as $0 \to 3$ and $0 \to 4$, so 0 is also transient. It is easy to see that 4 and 5 form a closed class and hence they are recurrent. Now it is easy to see that 2 and 3 form a class which is not closed, so they are transient. In summary, there are four classes: three transient classes $\{0\}$, $\{1\}$, and $\{2, 3\}$, and one recurrent class $\{4, 5\}$.

Example 4.9 Let X_n be a simple random walk on integers with step-up probability p. Recall $X_n = X_0 + \xi_1 + \cdots + \xi_n$, where ξ_n are iid with $P(\xi = 1) = p$ and $P(\xi = -1) = 1 - p$. By the SLLN,

$$\frac{X_n}{n} = \frac{X_0}{n} + \frac{\xi_1 + \cdots + \xi_n}{n} \to E(\xi) = 2p - 1$$

as $n \to \infty$. If $p \neq 1/2$, then the limit of X_n/n is nonzero, which implies that $X_n \to \pm\infty$ and hence all states are transient.

The case of a symmetric random walk with $p = 1/2$ is a little more difficult. It is shown in Example 1.7 that if $p = 1/2$, then $P_0(\tau_1 < \infty) = 1$, that is, with probability 1, the random walk will reach 1 from 0. Reversing the direction shows that with probability 1, the random walk will also reach 0 from 1. This implies that with probability 1, the random walk will return to 0, and hence 0 is a recurrent state. Since there is only one class, all states are recurrent for $p = 1/2$.

Example 4.10 For the simple branching process X_n (in Example 4.3) with offspring distribution $(p_0, p_1, p_2, \ldots)$, the state 0 is obviously an absorbing state and hence is recurrent. We will show that if $p_1 < 1$, then all nonzero states are transient.

To show this, first assume $p_0 > 0$. Then for $i > 0$,

$$P_i(\tau_i < \infty) \leq P_i(\text{at least 1 offspring}) = 1 - p_0^i < 1,$$

which implies the transience of i. If $p_0 = 0$, then X_n is nondecreasing in n and hence

$$P_i(\tau_i < \infty) = P(X_1 = i \mid X_0 = i) = p_1^i < 1.$$

This proves the transience of any $i > 0$ under $p_1 < 1$.

Because the number of visits to $i > 0$ is finite, it follows that with probability 1, either $X_n = 0$ for some $n > 0$ or $\lim_{n \to \infty} X_n = \infty$.

Let $\pi = P_1(X_n = 0$ for some $n > 0)$, the extinction probability given the population started with 1 individual. Let ξ be the number of offsprings from one individual. The pgf of ξ is $P(s) = E(s^\xi) = \sum_{i=0}^{\infty} p_i s^i$ and its mean is $\mu = P'(1-)$.

Theorem 4.11 *If $\mu \leq 1$, then $\pi = 1$ (sure extinction). On the other hand, if $\mu > 1$, then $\pi < 1$ is the unique solution of the equation $\pi = P(\pi)$ for $\pi \in [0, 1)$.*

Proof: Let $P_n(s)$ be the pgf of X_n given $X_0 = 1$. Then $P_1(s) = P(s)$. Note that $X_{n+1} = \sum_{j=1}^{X_n} \xi_j$, where ξ_j are iid with the same distribution as ξ and independent of X_n. By conditioning on X_n,

$$
\begin{aligned}
P_{n+1}(s) &= E[P(s)^{X_n}] = P_n(P(s)) = P_{n-1}(P(P(s))) \\
&= \cdots = P(P_n(s)). \tag{4.23}
\end{aligned}
$$

Because $\lim_{n \to \infty} P_n(0) = \lim_{n \to \infty} P(X_n = 0) = P(X_n = 0$ for some $n > 0) = \pi$, it follows from (4.23) that π is a solution of $\pi = P(\pi)$. Because $P'(s) \geq 0$ and $P''(s) \geq 0$ for $0 < s < 1$, the graph of $P(s)$, connecting $(0, p_0)$ and $(1, 1)$, is increasing and convex with slope μ at $(1, 1)$, see Figure 4.1, it is easy to see that if $\mu \leq 1$, then the only solution of equation $t = P(t)$ on $[0, 1]$ is $\pi = 1$, and if $\mu > 1$, then it has a unique solution in $[0, 1)$. $\diamondsuit$

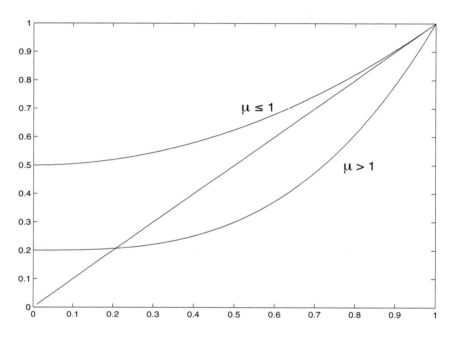

FIGURE 4.1: Graphs of $P(s)$.

Exercise 4.10 Consider an MC on states $0, 1, 2, 3, 4, 5$ with the transition probability matrix given below

$$
\mathbf{P} = \begin{bmatrix}
0.3 & 0 & 0.7 & 0 & 0 & 0 \\
0 & 0.3 & 0 & 0.7 & 0 & 0 \\
0.7 & 0 & 0.3 & 0 & 0 & 0 \\
0 & 0.2 & 0 & 0.8 & 0 & 0 \\
0.3 & 0.2 & 0 & 0 & 0.3 & 0.2 \\
0.1 & 0.2 & 0.2 & 0.1 & 0.2 & 0.2
\end{bmatrix}.
$$

Identify its classes and determine which classes are recurrent and which are transient.

Exercise 4.11 Assume a simple branching process X_n with the off-spring distribution given by $p_0 = 0.3$, $p_1 = 0.3$, and $p_2 = 0.4$.
(a) Find the extinction probability π given $X_0 = 1$.
(b) Find the probability that population becomes extinct before or at generation $n = 3$.

Exercise 4.12 In a simple branching process X_n with offspring distribution given by $p_0 = p$ and $p_1 = 1 - p$ $(0 < p < 1)$, let τ be the first time $n > 0$ when $X_n = 0$. For any integers $i > 0$ and $n > 0$, find $P(\tau = n \mid X_0 = i)$ in terms of p.

Exercise 4.13 For any two transient states i and j, show $E_i(N_j) = f_{ij}/(1 - f_{jj})$. Note that $f_{jj} < 1$ because j is transient.

4.5   Periodicity, class property, and positive recurrence

Period: The period of a state i is defined by

$$d(i) = \text{greatest common divisor of } \{n \geq 1; \ p_{ii}^{(n)} > 0\}. \qquad (4.24)$$

It is clear that if $p_{ii} > 0$, then $d(i) = 1$, but even when $p_{ii} = 0$, $d(i)$ may still be equal to 1. On the other hand, if $p_{ii}^{(n)} > 0$ only for even n, then $d(i) \geq 2$. For example, all states in a simple random walk have period 2, but for a simple random walk with absorbing boundary at 0 and N, states 0 and N have period 1, and other states have period 2.

Class properties By the following theorem, the recurrence, transience, and period are all class properties in the sense that all the states within the same class have the same period, and are either all recurrent or all transient. Therefore, for an irreducible MC, it makes sense to say it is recurrent, transient, or it has a certain period. An irreducible MC is called aperiodic if its period is one.

Theorem 4.12 *Assume $i \leftrightarrow j$. Then $d(i) = d(j)$. Moreover, if i is recurrent, then so is j.*

Proof: Choose n and m such that $p_{ij}^{(m)} > 0$ and $p_{ji}^{(n)} > 0$. Let $p_{ii}^{(s)} > 0$ and note that $d(i)$ is the largest integer which divides all such s. Since

$$p_{jj}^{(n+m)} \geq p_{ji}^{(n)} p_{ij}^{(m)} > 0 \quad \text{and} \quad p_{jj}^{(n+m+s)} \geq p_{ji}^{(n)} p_{ii}^{(s)} p_{ij}^{(m)} > 0,$$

it follows that $d(j)$ divides all s and hence $d(j) \le d(i)$. Reversing the argument shows $d(i) \le d(j)$, and hence $d(i) = d(j)$.

To prove j is recurrent if i is so, we need the following simple recurrence criterion.

Lemma 4.13 *A state i is recurrent if and only if $\sum_{n=1}^{\infty} p_{ii}^{(n)} = \infty$.*

Note that $\sum_{n=1}^{\infty} p_{ii}^{(n)} = \sum_{n=1}^{\infty} P_i(X_n = i) = E[\#$ of returns to $i \mid X_0 = i]$. By Theorem 4.7, this is ∞ if and only if i is recurrent. The lemma is proved.

Now assume i is recurrent and $i \leftrightarrow j$. Let n and m be chosen as before. For any $s > 0$, $p_{jj}^{(n+m+s)} \ge p_{ji}^{(n)} p_{ii}^{(s)} p_{ij}^{(m)}$. The recurrence of i implies $\sum_s p_{ii}^{(s)} = \infty$, $\sum_k p_{jj}^{(k)} \ge \sum_s p_{jj}^{(n+m+s)} = \infty$, and hence the recurrence of j. $\diamond$

Theorem 4.14 *If i is recurrent and $i \to j$, then $j \to i$ and hence j is also recurrent. Moreover, $f_{ij} = 1$.*

Proof: Since $i \to j$, if i is not accessible from j, then the return to i would have a probability < 1, which contradicts the recurrence of i. Therefore, $j \to i$, and by Theorem 4.12, j is recurrent. Since the MC will go from j to i with positive probability, if $f_{ij} < 1$, then the probability of returning to j would be less than 1, which contradicts the recurrence of j. $\diamond$

Positive or null recurrence: A recurrent state i is called positive recurrent if the mean return time is finite, that is, if $E_i(\tau_i) < \infty$. Otherwise, it is called null recurrent.

An MC is called positive or null recurrent if all its states are so. It will be shown later that the positive recurrence and null recurrence are also class properties, thus if one state of an irreducible MC is positive or null recurrent, then so is the MC.

Example 4.15 Recall that a symmetric simple random walk on all integers is shown in Example 4.9 to be recurrent. In Example 1.7, it is shown that $E_0(\tau_1) = \infty$. This implies that $E_{-1}(\tau_0) = \infty$ and, by symmetry, $E_1(\tau_0) = \infty$. Conditioning on the first step, we obtain

$$E_0(\tau_0) = \frac{1}{2} E_1(\tau_0) + \frac{1}{2} E_{-1}(\tau_0) = \infty.$$

This shows that the symmetric simple random walk is null recurrent.

Theorem 4.16 *A closed class with finitely many states is positive recurrent. Consequently, an irreducible MC with finitely many states is positive recurrent.*

Proof: It is shown in §4.4 that a closed class with finitely many states must be recurrent. Let i be a state in the class. We want to show $E_i(\tau_i) < \infty$. By Theorem 4.14, $f_{ji} = 1$ for any state j in the class, then there are an integer $N > 0$ and a real number $r > 0$ such that $P_j(\tau_i \leq N) \geq r$ for all j (here the finite number of states is used). Then $P_j(\tau_i > N) \leq 1 - r < 1$ for any j and hence by the Markov property at times kN for $k = 1, 2, 3, \ldots$, $P_j(\tau_i > kN) \leq (1-r)^k$ (see Exercise 4.15). It follows that

$$
\begin{aligned}
E_i(\tau_i) &= \sum_k E_i[\tau_i \mid (k-1)N < \tau_i \leq kN] P_i[(k-1)N < \tau_i \leq kN] \\
&\leq \sum_k kN P[\tau_i > (k-1)N] \leq N \sum_k k(1-r)^{k-1} < \infty. \ \Diamond
\end{aligned}
$$

Exercise 4.14 Let X_n be an irreducible MC with transition probabilities p_{ij} and period $d > 1$.
(a) Show that for any two states i and j, there is an integer $k = k(i, j)$ with $0 \leq k < d$ such that $p_{ij}^{(m)} = 0$ for any integer $m > 0$ unless $m = k + nd$ for some integer $n \geq 0$.
(b) Let S be the state space and fix $o \in S$. Show that S is divided into d nonempty disjoint subsets F_α for $\alpha = 0, 1, 2, \ldots, d-1$, where F_α is the set of states j with $k(o, j) = \alpha$.
(c) For $i \in F_\alpha$ and $j \in F_\beta$ with $\alpha, \beta = 0, 1, 2, \ldots, d-1$, show that $k(i, j) = \beta - \alpha$ if $\alpha \leq \beta$, and $k(i, j) = d - \alpha + \beta$ if $\alpha > \beta$. In particular, if i and j belong to the same F_α, then $d(i, j) = 0$, and hence $p_{ij}^{(m)} = 0$ for any integer $m > 0$ unless m is a multiple of d.
(d) Show that the MC X_n restricted to each F_α and to the time index set $\{0, d, 2d, 3d, \ldots\}$ is an aperiodic MC. More precisely, show that if the distribution of X_0 is concentrated on F_α, then $X_n^* = X_{nd}$ is an irreducible and aperiodic MC in the state space F_α with transition probabilities $p_{ij}^* = p_{ij}^{(d)}$.
(e) Show that X_n^* is positive recurrent if X_n is.

Exercise 4.15 For a general MC, suppose for fixed $b > 0$ and $N > 0$, $P_j(\tau_i > N) \leq b$ for any two states i and j in a closed class. Show that $P_j(\tau_i > kN) \leq b^k$ for $k = 1, 2, 3, \ldots$.

4.6 Expected hitting time and hitting probability

Hitting time of a set: Let X_n be an MC and let A be a set of its states. The hitting time of A is

$$\tau = \inf\{n > 0; \ X_n \in A\}. \tag{4.25}$$

This is the first time $n > 0$ when X_n is in A and is set to ∞ if $X_n \notin A$ for all $n > 0$.

Fix $i \in A^c$ (the complement of A) and $k \in A$. We are interested to find

$$v_i = E_i(\tau), \tag{4.26}$$

the mean time to hit the set A from i, and

$$u_i = P_i(X_\tau = k), \tag{4.27}$$

the probability of hitting k when A is hit.

Because we are only concerned with the behaviors of the MC up to the time when it enters A, we may and will assume that all the states in A are absorbing. Then $v_j = 0$ for $j \in A$. We will also assume that A^c consists of finitely many states from which A is accessible. Then the states in A^c are transient.

Finiteness of v_i: We note that starting from a transient state i, the mean time to hit A is finite, that is, $v_i = E_i(\tau) < \infty$. This follows from the fact that from each transient state, the expected number of returns is finite and there are only finitely many transient states.

Equations for u_i and v_i: By the first step analysis, that is, by conditioning on the first transition step of MC, we obtain

$$u_i = p_{ik} + \sum_{j \in A^c} p_{ij} u_j, \quad i \in A^c, \tag{4.28}$$

$$v_i = 1 + \sum_{j \in A^c} p_{ij} v_j, \quad i \in A^c. \tag{4.29}$$

Theorem 4.17 *The two systems of equations (4.28) and (4.29) have unique solutions u_i and v_i, $i \in A^c$.*

Proof: Let $\mathbf{Q}$ be the matrix $\{p_{ij}\}_{i,j \in A^c}$, a sub-matrix of the transition probability matrix $\mathbf{P}$, and let $\mathbf{u}$, $\mathbf{v}$, and $\mathbf{p}_k$ be, respectively, the column vectors containing u_i, v_i, and p_{ik} for $i \in A^c$. Then (4.28) and (4.29) may be written more concisely as

$$(\mathbf{I} - \mathbf{Q})\mathbf{u} = \mathbf{p}_k \quad \text{and} \quad (\mathbf{I} - \mathbf{Q})\mathbf{v} = \mathbf{1},$$

where $\mathbf{I}$ is the identity matrix and $\mathbf{1}$ is the column vector of 1's. The matrix $(\mathbf{I} - \mathbf{Q})$ has an inverse $(\mathbf{I} - \mathbf{Q})^{-1} = \sum_{n=0}^{\infty} \mathbf{Q}^n$ (the matrix series converges because its (i, j)-entry is the mean number of visits to a transient state j given starting at i, which is finite by Exercise 4.13). Therefore, (4.28) and (4.29) may be solved uniquely. $\diamond$

Example 4.18 A rat is put in a maze of nine rooms numbered 1 through 9 shown below.

1	2	3
4	5	6
7	8	9

Assume rooms 3 and 7 are traps; the former contains food and the latter contains an electric shock. In all the other rooms, the rat will stay 1 unit time and then select an adjacent room (across a wall) randomly. For any $i = 1, 2, \ldots, 9$, find
(a) $P(\text{rat finds food} \mid \text{it starts in room i})$, and
(b) the mean time until rat enters a trap given it starts in room i.

Solution: Let X_n be the room number at time n. Then X_n is a Markov chain with nine states $1, 2, \ldots, 9$, among which 3 and 7 are absorbing. We want to find u_i in (a) and v_i in (b) with state $k = 3$ as the desired absorbing state. Note that $u_3 = 1$ and $u_7 = v_3 = v_7 = 0$. According to (4.28) and (4.29), we obtain seven equations for u_i and seven equations for v_i, $i \neq 3, 7$. It is in fact often easier to obtain these equations by directly applying the first step analysis. Moreover, using symmetry, the number of equations may be greatly reduced. Indeed, $u_1 = u_5 = u_9 = 1/2$,

$$u_6 = u_2 = \frac{1}{3}u_1 + \frac{1}{3}u_5 + \frac{1}{3}u_3 = \frac{2}{3}$$

and

$$u_8 = u_4 = \frac{1}{3}u_1 + \frac{1}{3}u_5 + \frac{1}{3}u_7 = \frac{1}{3}.$$

Also by symmetry, $v_1 = v_9$, $v_2 = v_4 = v_6 = v_8$. We have

$$
\begin{cases}
v_1 &= 1 + \frac{1}{2}v_2 + \frac{1}{2}v_4 = 1 + v_2 \\
v_2 &= 1 + \frac{1}{3}v_1 + \frac{1}{3}v_3 + \frac{1}{3}v_5 = 1 + \frac{1}{3}v_1 + \frac{1}{3}v_5 \\
v_5 &= 1 + \frac{1}{4}v_2 + \frac{1}{4}v_4 + \frac{1}{4}v_6 + \frac{1}{4}v_8 = 1 + v_2.
\end{cases}
$$

Solving the above three equations, we obtain $v_1 = v_5 = v_9 = 6$ and $v_2 = v_4 = v_6 = v_8 = 5$.

Example 4.19 A fair coin is tossed repeatedly.
(a) Find the expected number of tosses until HTH appears.
(b) Find P(HTH appears before HHH).

Solution: (a) Consider a Markov chain with four states: 0 (no part of HTH appears), 1 (H but not HTH), 2 (HT), and 3 (HTH) with state 3 being absorbing. The transition probabilities are $P_{00} = P_{01} = 1/2$, $P_{11} = P_{12} = 1/2$, $P_{20} = P_{23} = 1/2$, $P_{33} = 1$, and all other $P_{ij} = 0$. This is depicted below,

$$\hookrightarrow \ (0) \ \rightarrow \ (1) \ \leftrightarrow \ \rightarrow \ (2) \ \rightarrow \ (3) \ \overset{1}{\hookleftarrow} \ ,$$

where all arrows have the same transition probability $1/2$ except the last one, which is indicated to have probability 1. By the first step analysis,

$$v_0 = 1 + (1/2)v_0 + (1/2)v_1, \quad v_1 = 1 + (1/2)v_1 + (1/2)v_2, \quad v_2 = 1 + (1/2)v_0.$$

Solve to get $v_0 = 10$ ($v_1 = 8$ and $v_2 = 6$).
(b) Now consider a Markov chain with six states: 0 (no part of HTH and HHH), 1 (H but not HTH and HH), 2 (HT), 3 (HTH, absorbing), 4 (HH but not HHH), 5 (HHH, absorbing). The transition probabilities are $P_{00} = P_{01} = P_{12} = P_{14} = P_{20} = P_{23} = P_{42} = P_{45} = 1/2$, $P_{33} = P_{55} = 1$, and all other $P_{ij} = 0$. This is depicted below:

$$
\begin{array}{c}
\swarrow \ \leftarrow \ \nwarrow \\
\hookrightarrow \ (0) \ \rightarrow \ (1) \ \rightarrow \ (2) \ \rightarrow \ (3) \ \overset{1}{\hookleftarrow} \\
\downarrow \quad \nearrow \\
(4) \ \rightarrow \ (5) \ \overset{1}{\hookleftarrow},
\end{array}
$$

where all arrows are associated with the same transition probability

1/2 except the two indicated as to probability 1. With state 3 being the desired absorbing state, the first step analysis leads to

$$u_0 = (1/2)u_0 + (1/2)u_1, \quad u_1 = (1/2)u_2 + (1/2)u_4,$$
$$u_2 = (1/2)u_0 + 1/2, \quad u_4 = (1/2)u_2.$$

Solve these equations to get the desired probability $u_0 = 3/5$ ($u_1 = 3/5$, $u_2 = 4/5$, $u_4 = 2/5$).

Example 4.20 (Gambler's ruin problem): Let X_n be a simple random walk on $\{0, 1, 2, \ldots, N\}$ with absorbing boundary and 1-step transition probabilities $p_{i\,i+1} = p$ and $p_{i\,i-1} = q = 1 - p$ for $1 < i < N$, $p_{00} = p_{NN} = 1$, and all other $p_{ij} = 0$. Assume $0 < p < 1$.

This process describes a gambler's fortune at time n in a series of \$1 bets with p being the probability of winning the bet. The game stops when his fortune reaches 0 or N. Then given initial fortune i, v_i is the expected time of the game, and u_i is the probability of gambler's ruin when $k = 0$ is taken as the "desired" absorbing state.

From (4.28) and (4.29), for $0 < i < N$,

$$u_i = qu_{i-1} + pu_{i+1} \quad \text{and} \quad v_i = 1 + qv_{i-1} + pv_{i+1}.$$

Noting $p + q = 1$, one gets

$$u_{i+1} - u_i = \frac{q}{p}(u_i - u_{i-1}) = \cdots = (\frac{q}{p})^i(u_1 - u_0)$$

and

$$v_{i+1} - v_i = \frac{q}{p}(v_i - v_{i-1}) - \frac{1}{p} = \cdots = (\frac{q}{p})^i(v_1 - v_0) - \frac{1}{p}[1 + \frac{q}{p} + \cdots + (\frac{q}{p})^{i-1}].$$

Using $u_0 = 1$, $u_N = v_0 = v_1 = 0$, we may get with some effort that for $0 \le i \le N$,

$$u_i = \begin{cases} \frac{(q/p)^i - (q/p)^N}{1 - (q/p)^N} & \text{if } p \ne q \\ \frac{N-i}{N} & \text{if } p = q. \end{cases} \quad (4.30)$$

$$v_i = \begin{cases} \frac{1}{p-q}[N\frac{1-(q/p)^i}{1-(q/p)^N} - i] & \text{if } p \ne q \\ \frac{1}{2p}i(N - i) & \text{if } p = q. \end{cases} \quad (4.31)$$

We may consider a slightly generalized version of a simple random walk on $\{0, 1, 2, \ldots, N\}$ with absorbing boundary by allowing $p_{ii} = r$ for $1 \le i \le N - 1$, where $r > 0$ is the probability of breakeven in a bet so that $p + q + r = 1$. Then it can be shown that the expressions in (4.30) and (4.31) still hold.

Exercise 4.16 As in Example 4.18, a rat is put in a maze of nine rooms, but now assume rooms 7 and 9 are traps, the former contains food and the latter contains an electric shock. For $i = 1, 2, 3, 4, 5, 6$, and 8, find
(a) $u_i = P(\text{rat finds food} \mid \text{it starts in room i})$, and
(b) the mean number v_i of steps until rat enters a trap given it starts in room i.

Exercise 4.17 A fair die is rolled repeatedly.
(a) Find the expected number of rolls until pattern 666 appears.
(b) Find the expected number of rolls until 123 appears.
(c) Find $P(123 \text{ appears before } 111)$.

Exercise 4.18 A stock price stays constant for 1 unit of time and then it may either go up 1 unit with probability 0.4 or go down 1 unit with probability 0.6. Suppose the current price is 50. Find the probability that the price will rise up to 60 before falling down to 40.

Exercise 4.19 Under the general setup of this section, let $P_i(s)$ be the pgf (probability generating function) of τ, the hitting time of A.
(a) Using the first step analysis, derive a system of equations for $P_i(s)$ from which one may solve for $P_i(s)$.
(b) In Example 4.18, find the pgf $P_i(s)$ for the time τ when the rat enters a trap given it starts in room $i = 1, 2, 4, 5, 6, 8, 9$, and from which to find $\text{Var}(\tau)$.

4.7 Stationary distribution

Definition of stationary distribution: Let X_n be an MC with state space S and transition probabilities p_{ij}. A distribution on S is a set of numbers π_j, one for each state $j \in S$, such that $\pi_j \geq 0$ and $\sum_j \pi_j = 1$. It is called a stationary distribution of the MC if for any $j \in S$,

$$\pi_j = \sum_i \pi_i p_{ij}. \qquad (4.32)$$

It is clear that if $\{\pi_j\}$ is a stationary distribution, then by (4.15),

$$\pi_j = \sum_i \pi_i p_{ij}^{(n)} \quad \text{for all } j \text{ and } n \geq 1. \qquad (4.33)$$

It follows that if the MC X_n starts with a stationary distribution π_j as initial distribution, then X_n has the same distribution at all times n. In fact, X_n is a stationary process as defined in §3.7, that is, for any $k > 0$, the joint distribution of $X_{n+1}, X_{n+2}, \ldots, X_{n+k}$ is independent of n, because

$$P(X_{n+1} = j_1, X_{n+2} = j_2, \ldots, X_{n+k} = j_k)$$
$$= \sum_i \pi_i p_{ij_1}^{(n+1)} p_{j_1 j_2} \cdots p_{j_{k-1} j_k} = \pi_{j_1} p_{j_1 j_2} \cdots p_{j_{k-1} j_k}.$$

One may think of a distribution as a unit mass distributed in the state space. It is stationary as defined by equation (4.32) means that the mass at state j (the left-hand side of equation) is equal to the mass transferred to state j in one step (the right-hand side of equation).

Positivity of stationary distribution: Let $\{\pi_j\}$ be a stationary distribution of an irreducible MC. Then all $\pi_j > 0$. This is an easy consequence of (4.33).

Main results of stationary distribution: Let X_n be an irreducible MC. It will be shown in the next section that if X_n is positive recurrent, then it has a unique stationary distribution $\{\pi_j\}$. On the other hand, if it is either transient or null recurrent, then it has no stationary distribution. Moreover, in the positive recurrent case, π_j is the long-run fraction of times when the MC is in state j, that is,

$$\pi_j = \lim_{n \to \infty} \frac{\text{number of } k \text{ such that } X_k = j \text{ for } 1 \le k \le n}{n} \quad a.s. \quad (4.34)$$

Furthermore, if in addition the MC is aperiodic, then π_j is also the long-run (or limiting) probability that the MC is in state j, that is,

$$\pi_j = \lim_{n \to \infty} P(X_n = j), \quad (4.35)$$

regardless of the initial distribution of the MC. In particular, $\pi_j = \lim_{n \to \infty} p_{ij}^{(n)}$ for any two states i and j.

Theorem 4.21 *If the MC is irreducible and has finitely many states, then the system of equations in (4.32) together with $\sum_j \pi_j = 1$ has a unique solution $\{\pi_j\}$, which is necessarily the stationary distribution.*

Proof: By Theorem 4.16, the MC is positive recurrent and hence has a unique positive solution $\{\pi_j\}$ as mentioned above. If the equations have another solution $\{\pi'_j\}$, then by choosing a large enough $a > 0$, $\tilde{\pi}_j = a\pi_j + \pi'_j \geq 0$ and $\tilde{\pi}_j = \sum_i \tilde{\pi}_i p_{ij}$ for all j. Thus, we may find a constant $b > 0$ such that $\{b\tilde{\pi}_j\}$ becomes a stationary distribution. By the uniqueness of the stationary distribution, $b\tilde{\pi}_j = \pi_j$ and hence π'_j is proportional to π_j. This implies $\pi'_j = \pi_j$. ◊

Determination of stationary distribution: The stationary distribution $\{\pi_j\}$ may be obtained by solving the equations in (4.32) together with $\sum_j \pi_j = 1$. Note that the equations in (4.32) are redundant because they sum to an identity, and so one of them may be dropped in solving for π_j. Theorem 4.22 below provides an alternative way to find π_j.

Theorem 4.22 *Let X_n be an irreducible MC with a finite number m states. Then the unique stationary distribution $\pi = (\pi_1 \ \pi_2 \ \ldots \ \pi_m)$, regarded as a row vector, is given by*

$$\pi = 1_{1 \times m}(I - \mathbf{P} + 1_{m \times m})^{-1}, \tag{4.36}$$

where I is the identity matrix and $1_{k \times m}$ is a $k \times m$ matrix of 1's, and the inverse matrix in (4.36) exists.

Proof: Because the MC X_n is positive recurrent, it has a unique stationary distribution π satisfying $\pi(I - \mathbf{P}) = 0$ and $\pi 1_{m \times m} = 1_{1 \times m}$. Adding up, we obtain a single equation

$$\pi(I - \mathbf{P} + 1_{m \times m}) = 1_{1 \times m}. \tag{4.37}$$

Conversely, if π is a solution of equation (4.37), multiplying both sides of (4.37) by $1_{m \times 1}$ on the right and using $\mathbf{P}1_{m \times 1} = 1_{m \times 1}$ yields $m \sum_j \pi_j = m$. Then $\sum_j \pi_j = 1$ and $\pi 1_{m \times m} = 1_{1 \times m}$, and by (4.37), π satisfies (4.32) and hence is a stationary distribution. By the uniqueness of the stationary distribution, the equation (4.37) has a unique solution. This implies the invertibility of its coefficient matrix $(I - \mathbf{P} + 1_{m \times m})$. ◊

Example 4.23 Consider an MC of three states with transition matrix

$$\mathbf{P} = \begin{bmatrix} 0.2 & 0.3 & 0.5 \\ 0.3 & 0.4 & 0.3 \\ 0.6 & 0.3 & 0.1 \end{bmatrix}.$$

By (4.36), its stationary distribution $\pi = (\pi_1 \; \pi_2 \; \pi_3)$ is given by

$$(1\;1\;1)(I - \mathbf{P} + \mathbf{1}_{3\times3})^{-1} = (1\;1\;1) \begin{bmatrix} 1.8 & 0.7 & 0.5 \\ 0.7 & 1.6 & 0.7 \\ 0.4 & 0.7 & 1.9 \end{bmatrix}^{-1}$$

$$= (0.3571\;0.3333\;0.3095) \;\; (\text{by MATLAB}^{\circledR}).$$

Example 4.24 (Reflecting random walk): The reflecting random walk on $\{0, 1, 2, \ldots, N\}$ given in Example 4.2, with step-up probability p $(0 < p < 1)$, is an irreducible and positive recurrent MC. To find the stationary distribution π_j, we write down the equations (4.32) as

$$\pi_0 = \pi_1 q, \quad \pi_1 = \pi_0 + \pi_2 q, \quad \pi_2 = \pi_1 p + \pi_3 q, \quad \ldots, \quad (q = 1 - p)$$
$$\pi_{N-2} = \pi_{N-3} p + \pi_{N-1} q, \quad \pi_{N-1} = \pi_{N-2} p + \pi_N, \quad \pi_N = \pi_{N-1} p.$$

Solving recursively but not using the last equation, we obtain

$$\pi_1 = q^{-1}\pi_0, \quad \pi_2 = pq^{-2}\pi_0, \quad \pi_3 = p^2 q^{-3}\pi_0, \quad \ldots$$
$$\pi_{N-1} = p^{N-2}q^{-(N-1)}\pi_0, \quad \pi_N = p^{N-1}q^{-(N-1)}\pi_0.$$

This determines all π_j except π_0. By $\sum_{j=0}^{N} \pi_j = 1$, π_0 is determined by

$$\pi_0 = \frac{1}{1 + q^{-1} + pq^{-2} + \cdots + p^{N-2}q^{-(N-1)} + p^{N-1}q^{-(N-1)}}$$

$$= \begin{cases} 1/(2N), & \text{if } p = 1/2 \\ 1/[1 + (q^{N-1} - p^{N-1})/(q^N - pq^{N-1}) + (p/q)^{N-1}], & \text{if } p \neq 1/2. \end{cases}$$

In particular, for the symmetric reflecting random walk with $p = 1/2$, $\pi_0 = \pi_N = 1/(2N)$ and $\pi_i = 1/N$ for $1 \leq i \leq (N - 1)$.

Example 4.25 A fair coin is tossed repeatedly. Find the long-run fraction of tosses when the pattern HTH appears. Solve the problem by counting overlapped patterns (so HTHTH is counted as two appearances of HTH). In Exercise 4.22, the same problem is solved when not counting a pattern that overlaps a preceding one.

Solution: Allowing the pattern to overlap, the Markov chain is as in Example 4.19 (a) but state 3 (HTH) is no longer absorbing with $p_{31} = p_{32} = 1/2$. Then solving

$$\pi_0 = (1/2)\pi_0 + (1/2)\pi_2, \quad \pi_2 = (1/2)\pi_1 + (1/2)\pi_3, \quad \pi_3 = (1/2)\pi_2,$$

together with $\pi_0 + \pi_1 + \pi_2 + \pi_3 = 1$ yields $\pi_0 = \pi_2 = 1/4$, $\pi_1 = 3/8$, and $\pi_3 = 1/8$, which is the long run-fraction of tosses when HTH appears.

Note that π_3 is just the probability of obtaining HTH in three consecutive tosses, that is, $(1/2)^3$. In fact, $\pi_3 = (1/2)^3$ for any pattern of three consecutive faces. It may be justified using the limiting probability $\pi_j = \lim_{n\to\infty} p_{ij}^{(n)}$ noting $p_{i3}^{(n)} = (1/2)^3$ for $n \geq 3$.

Example 4.26 (Discrete queuing model): In the discrete queuing model Example 4.5, the MC X_n is the number of customers at the end of the nth period. It has infinitely many states $0, 1, 2, \ldots$. Let $a_k = P(A = k)$, where A is the number of arrivals during a service period. Assume $a_k > 0$ for $k = 0$ and some $k > 1$. Then X_n is irreducible. The stationary distribution π_j, if exists, satisfies the equations

$$
\begin{aligned}
\pi_j &= \pi_0 a_j + \pi_1 a_j + \pi_2 a_{j-1} + \ldots + \pi_j a_1 + \pi_{j+1} a_0 \\
&= \pi_0 a_j + \sum_{i=1}^{j+1} \pi_i a_{j+1-i} \quad \text{for } j \geq 0.
\end{aligned}
\tag{4.38}
$$

In particular,

$$
\pi_0 = \pi_0 a_0 + \pi_1 a_0, \qquad \pi_1 = \pi_0 a_1 + \pi_1 a_1 + \pi_2 a_0, \quad \ldots.
$$

The equation (4.38) is an recursive relation from which we may determine $\pi_1, \pi_2, \pi_3, \ldots$ successively if π_0 is known. Let

$$
\rho = E(A) = \sum_{j=0}^{\infty} j a_j
\tag{4.39}
$$

be the mean number of arrivals in a service period. By Theorem 4.27 below, if $\rho < 1$, then the stationary distribution exists with $\pi_0 = 1 - \rho$.

Note that if X_n is the total queue length of an $M/G/1$ system at the end of the nth service (see Example 4.5), then (4.38) is just (3.50) in §3.8, and $\rho = E(A) = E(\lambda Y)$ is the traffic intensity as defined in §3.3, where λ is the Poisson arrival rate and Y is a generic service time.

Theorem 4.27 *The irreducible MC of the discrete queuing model above has a stationary distribution (that is, is positive recurrent) if and only if $\rho < 1$, and in this case, $\pi_0 = 1 - \rho$.*

Proof: Let $b_j = \sum_{i=j}^{\infty} a_i = P(A \geq j)$. We will show (4.38) implies

$$\pi_1 a_0 = \pi_0 b_1 \quad \text{and} \quad \pi_j a_0 = \pi_0 b_j + \sum_{i=1}^{j-1} \pi_i b_{j+1-i} \quad \text{for } j \geq 2. \quad (4.40)$$

The first few equations in (4.40) are listed below.

$$\begin{aligned}
\pi_1 a_0 &= \pi_0 b_1 \\
\pi_2 a_0 &= \pi_0 b_2 + \pi_1 b_2 \\
\pi_3 a_0 &= \pi_0 b_3 + \pi_1 b_3 + \pi_2 b_2 \\
\pi_4 a_0 &= \pi_0 b_4 + \pi_1 b_4 + \pi_2 b_3 + \pi_3 b_2, \quad \cdots
\end{aligned}$$

It is easy to verify directly that the first equation in (4.40) follows from (4.38) with $j = 0$, and then (4.40) with $j = 2$ follows from (4.38) with $j = 1$. Now assume (4.40) holds for a fixed j. Then by (4.38),

$$\begin{aligned}
\pi_{j+1} a_0 &= \pi_j (1 - a_1) - \left(\pi_0 a_j + \sum_{i=1}^{j-1} \pi_i a_{j+1-i} \right) \\
&= \pi_j a_0 + \pi_j b_2 - \left(\pi_0 a_j + \sum_{i=1}^{j-1} \pi_i a_{j+1-i} \right) \\
&= \pi_0 b_j + \sum_{i=1}^{j-1} \pi_i b_{j+1-i} + \pi_j b_2 - \left(\pi_0 a_j + \sum_{i=1}^{j-1} \pi_i a_{j+1-i} \right) \\
&= \pi_0 (b_j - a_j) + \sum_{i=1}^{j-1} \pi_i (b_{j+1-i} - a_{j+1-i}) + \pi_j b_2 \\
&= \pi_0 b_{j+1} + \sum_{i=1}^{j-1} \pi_i b_{j+2-i} + \pi_j b_2 = \pi_0 b_{j+1} + \sum_{i=1}^{j} \pi_i b_{j+2-i}.
\end{aligned}$$

This is (4.40) with j replaced by $j + 1$, and hence proves (4.40) from (4.38). See Exercise 4.23 for an alternative way to prove (4.40).

Let π_j be determined by (4.38) recursively with an arbitrarily fixed $\pi_0 > 0$. By (4.40), all $\pi_j \geq 0$. Summing (4.40) over all $j \geq 1$, using $\sum_{j=1}^{\infty} b_j = \rho$ and noting the pattern displayed below (4.40),

$$\begin{aligned}
a_0 \sum_{j=1}^{\infty} \pi_j &= \pi_0 \rho + \pi_1 (\rho - b_1) + \pi_2 (\rho - b_1) + \pi_3 (\rho - b_1) + \cdots \\
&= \pi_0 \rho + (\rho - b_1) \sum_{j=1}^{\infty} \pi_j.
\end{aligned}$$

Because $a_0 + b_1 = 1$, if $\sum_{j=1}^{\infty} \pi_j < \infty$, then the above equation may be written as $(1-\rho)\sum_{j=1}^{\infty} \pi_j = \pi_0 \rho$. We may suitably scale π_j to have $\sum_{j=1}^{\infty} \pi_j + \pi_0 = 1$. Then $(1-\rho)(1-\pi_0) = \pi_0 \rho$ and $\pi_0 = 1-\rho$. It is now clear that if $\rho \geq 1$, there is no stationary distribution. To show there is a stationary distribution when $\rho < 1$, it remains to show $\sum_{j=1}^{\infty} \pi_j < \infty$ with $\pi_0 > 0$ arbitrarily fixed. Summing the first N equations in (4.40),

$$a_0 \sum_{j=1}^{N} \pi_j = \pi_0 \sum_{j=1}^{N} b_j + \pi_1 \sum_{j=1}^{N} b_j + \pi_2 \sum_{j=2}^{N} b_j + \cdots + \pi_{N-1} b_2$$

$$\leq \pi_0 \sum_{j=1}^{N} b_j + \left(\sum_{j=2}^{N} b_j\right)\sum_{i=1}^{N} \pi_i.$$

Using $a_0 + b_1 = 1$, the above implies $(1-\sum_{i=1}^{N} b_i)\sum_{j=1}^{N} \pi_j \leq \pi_0 \sum_{j=1}^{N} b_j$. Because $\sum_{j=1}^{N} b_j \to \rho$ as $N \to \infty$ and $\rho < 1$, we obtain $\sum_{j=1}^{\infty} \pi_j \leq \pi_0 \rho/(1-\rho) < \infty$. $\Diamond$

Note on $\rho \geq 1$: A more detailed analysis will show that the MC in the discrete queuing model is null recurrent when $\rho = 1$ and transient when $\rho > 1$. See [9, Proposition 2.15.2].

Exercise 4.20 (a) For the inventory model with $a_n = 1/2^{n+1}$ for $n = 0, 1, 2, 3, \ldots$, and $s = 2$ and $S = 4$, find the long-run fraction of periods when restocking is required.
(b) Suppose each unit of merchandise kept per period incurs a cost of 1 and each replenishment incurs a fixed cost of 4 plus 2 per unit. Find the long-run cost per period.

Exercise 4.21 In Exercise 4.9 (a barber shop with 3 chairs and Poisson arrivals of rate 2), find the long-run fraction of times when the shop is empty.

Exercise 4.22 As in Example 4.25, a fair coin is tossed repeatedly. Find the long-run fraction of tosses when the pattern HTH appears, but not counting a pattern that overlaps a preceding one (so HTHTH is counted only as one appearance of HTH).

Exercise 4.23 Suppose the state space of a Markov chain is divided into two disjoint subsets A and B. Show a stationary distribution $\{\pi_j\}$ satisfies the following equation:

$$\sum_{i \in A} \sum_{j \in B} \pi_i p_{ij} = \sum_{j \in B} \sum_{i \in A} \pi_j p_{ji}.$$

This is called a balance equation because, thinking of $\{\pi_j\}$ as a distribution of unit mass among states, the left-hand side is the amount of mass transferred from A to B, and the right-hand side is that from B to A. Note that the equation (4.40) in Example 4.26 (discrete queuing model) is the balance equation with $A = \{0, 1, 2, \ldots, j-1\}$ and $B = \{j, j+1, \ldots\}$.

Exercise 4.24 Consider a simple random walk on nonnegative integers $0, 1, 2, \ldots$ with reflecting boundary at 0. Its transition probabilities are given by $p_{i\,i+1} = p$ and $p_{i-1\,i} = q = 1 - p$ for $i \geq 1$, $p_{01} = 1$, and all other $p_{ij} = 0$, for some $p \in (0, 1)$. It is easy to see that this is an irreducible Markov chain. Determine whether it is positive recurrent, null recurrent or transient, and find the stationary distribution if it is positive recurrent. The conclusion should depend on the value of p.

4.8 Limiting properties

Let X_n be a discrete time MC with state space S and transition probabilities p_{ij}. Recall that P_i is the conditional probability given $X_0 = i$ and τ_i is the hitting time of i.

Theorem 4.28 *If $i \in S$ is positive recurrent, then $\{\pi_j\}$ defined by*

$$\pi_j = \frac{E_i[\#\{n > 0; \ X_n = j \text{ for } n \leq \tau_i\}]}{E_i(\tau_i)}, \quad \text{for } j \in S, \qquad (4.41)$$

is a stationary distribution. Moreover, if $i \to j$, then $\pi_j > 0$.

Proof: It is easy to see that $\sum_j \pi_j = 1$. Let ν_j be the numerator in (4.41). Then for $j \neq i$,

$$\nu_j = \sum_{n=1}^{\infty} P_i(X_n = j, n \leq \tau_i)$$

$$= p_{ij} + \sum_{n=2}^{\infty} \sum_k P_i(X_n = j, n \leq \tau_i, X_{n-1} = k)$$

$$= p_{ij} + \sum_{n=2}^{\infty} \sum_k P_i(X_n = j \mid n \leq \tau_i, X_{n-1} = k) P_i(n \leq \tau_i, X_{n-1} = k)$$

$$= p_{ij} + \sum_{n=2}^{\infty} \sum_k P_i(X_n = j \mid [\tau_i \leq n - 1]^c \cap [X_{n-1} = k])$$
$$\times P_i(n \leq \tau_i, X_{n-1} = k)$$

$$= \nu_i p_{ij} + \sum_{n=2}^{\infty} \sum_k p_{kj} P_i(n \leq \tau_i, X_{n-1} = k)$$

(by Markov property and $\nu_i = 1$)

$$= \nu_i p_{ij} + \sum_{k \neq i} p_{kj} \nu_k = \sum_k \nu_k p_{kj}.$$

This shows that $\pi_j = \sum_k \pi_k p_{kj}$ for $j \neq i$. Summing over $j \neq i$ yields $1 - \pi_i = \sum_k \pi_k(1 - p_{ki})$, and hence $\pi_i = \sum_k \pi_k p_{ki}$. This shows that $\{\pi_j\}$ is a stationary distribution.

If $\pi_j = 0$, then from i, the probability of X_n hitting j before τ_i is zero. By the strong Markov property at time τ_i, X_n will never hit j and hence $i \to j$ is impossible. $\diamond$

Corollary 4.29 *In Theorem 4.28,*

$$E_i[\#\{n > 0; \quad X_n = j \text{ for } n \leq \tau_i\}] = \frac{\pi_j}{\pi_i},$$

that is, under P_i, the expected number of visits to j between two successive visits to i is equal to the ratio π_j/π_i.

Proof: This is an immediate consequence of Theorem 4.28. $\diamond$

Lemma 4.30 *Let τ_j^n, $n \geq 1$, be the successive hitting times of a recurrent state j and let $\sigma_n = \tau_j^n - \tau_j^{n-1}$ ($\tau_j^0 = 0$). Then for any state i in the same class as j, under P_i, σ_n are independent, and the distribution of σ_n for $n \geq 2$ is the same as that of $\tau_j = \tau_j^1$ under P_j.*

Proof: We will only prove the independence of $\tau = \tau_j$ and $\sigma = \tau_j^2 - \tau_j$ under P_i, and identify the distribution of σ. The general result is proved in the same way but involves a more complicated notation. By Theorem 4.14, $\tau_j < \infty$ a.s. under P_i. For any two integers $m, n \geq 0$,

$$P_i(\tau = n, \sigma = m) = P_i(\tau = n)P_i(\sigma = m \mid \tau = n, X_\tau = j)$$
$$= P_i(\tau = n)P_j(\tau = m) \quad \text{(by the strong Markov property at } \tau\text{)}.$$

Summing over n shows that $P_i(\sigma = m) = P_j(\tau = m)$. $\diamond$

An associated renewal reward process: Let i be a positive recurrent state and $i \to j$. Under P_i, considering a renewal reward process based on iid cycles of successive visits to i, with a unit reward for each visit to j, by its limiting property (Theorem 3.5), the long-run fraction of times when j is visited is $\pi_j > 0$ given in (4.41).

We may use the same reward, but based on the delayed renewal process of successive visits to j. By Theorem 4.14, j is recurrent, and then by Lemma 4.30, the cycle time σ_2 is finite with mean $E_j(\tau_j)$ under P_i. Then $\pi_j = 1/E_j(\tau_j)$, because $\pi_j > 0$, $E_j(\tau_j)$ is finite, and hence j is positive recurrent. This proves that the positive recurrence is a class property, and then so is the null recurrence. This fact is recorded below.

Theorem 4.31 *The null or positive recurrence is a class property, that is, if $i \leftrightarrow j$ and if i is null or positive recurrent, then so is j.*

Theorem 4.32 *(Long-run time average): Let*

$$N_j(m) = \#\{k > 0; \ X_k = j \text{ and } k \leq m\},$$

the number of visits to state j by time $m > 0$. Then for any two states i and j, a.s. under P_i,

$$\lim_{m \to \infty} \frac{N_j(m)}{m} = \begin{cases} 0, & \text{if } i \nrightarrow j \text{ or } j \text{ is not positive recurrent} \\ \pi_j, & \text{if } i \leftrightarrow j \text{ and positive recurrent,} \end{cases} \quad (4.42)$$

where π_j is the stationary distribution given in Theorem 4.28. Note that the limit in (4.42) is the long-run fraction of times when the MC is in state j. Moreover, $\pi_j = 1/E_j(\tau_j)$ when $i \leftrightarrow j$ are positive recurrent.

Proof: The case of positive recurrent $i \leftrightarrow j$ together with $\pi_j = 1/E_j(\tau_j)$ was proved in the earlier discussion of an associated renewal reward process. If $i \nrightarrow j$, then the MC will not visit j under P_i, and if j is transient, then it will visit j only finitely many times. In these two cases, the limit in (4.42) is equal to 0. Let σ_n be the successive inter-visit times to j as in Lemma 4.30. Then σ_n, $n \geq 2$, are iid under P_i. Note that if $n = N_j(m)$, then $N_j(m)/m \leq n/(\sigma_1 + \cdots + \sigma_n)$. If j is null recurrent, then $E_i(\sigma_2) = E_j(\tau_j) = \infty$ and by the SLLN, $\lim_{m \to \infty} N_j(m)/m \leq \lim_{n \to \infty} n/(\sigma_1 + \cdots + \sigma_n) = 1/E_i(\sigma_2) = 0$. This proves (4.42) for a null recurrent j. $\diamondsuit$

Corollary 4.33 *(Mean time average): In Theorem 4.32,*

$$\lim_{m \to \infty} \frac{1}{m} \sum_{n=1}^{m} p_{ij}^{(n)} = \begin{cases} 0, & \text{if } i \nrightarrow j \text{ or } j \text{ is not positive recurrent} \\ \pi_j, & \text{if } i \leftrightarrow j \text{ and positive recurrent.} \end{cases}$$

Proof: Applying E_i to (4.42), and noting that $E_i[N_j(m)] = \sum_{k=1}^{m} p_{ij}^{(k)}$ and $0 \leq N_j(m)/m \leq 1$, the above relation follows from the bounded convergence. $\diamondsuit$

Theorem 4.34 *(Existence and uniqueness of stationary distribution): Let X_n be an irreducible MC.*
(a) If X_n is either transient or null recurrent, then it has no stationary distribution.
(b) If X_n is positive recurrent, then it has a unique stationary distribution $\{\pi_j\}$ given by $\pi_j = 1/E_j(\tau_j)$.

Proof: Let b_j be a stationary distribution. Then $b_j = \sum_i b_i p_{ij}^{(n)}$ for any $n \geq 0$ and hence $b_j = \sum_i b_i (1/m) \sum_{n=1}^{m} p_{ij}^{(n)}$. Now let $m \to \infty$. By Corollary 4.33, if X_n is transient or null recurrent, then all $b_j = 0$, which is impossible, and hence there is no stationary distribution. If X_n is positive recurrent, then we get $b_j = \pi_j$. This proves the uniqueness of stationary distribution. Its existence is given by Theorem 4.28. In the above argument, taking limit under the summation sign is justified because $(1/m) \sum_{n=1}^{m} p_{ij}^{(n)} \leq 1$. The equality $\pi_j = 1/E_j(\tau_j)$ is already established in Theorem 4.32. $\diamondsuit$

Theorem 4.35 *(Limiting probabilities): Let X_n be an irreducible and positive recurrent MC with transition probabilities p_{ij} and stationary distribution $\{\pi_j\}$. If X_n is aperiodic, then for any two states i and j,*

$$\lim_{n \to \infty} p_{ij}^{(n)} = \pi_j. \tag{4.43}$$

If X_n is periodic with period d, then for any state j,

$$\lim_{n\to\infty} p_{jj}^{(nd)} = d\pi_j. \tag{4.44}$$

Proof: Let $X(t)$ be defined by $X(t) = X_n$ for $n \le t < n+1$. Then under P_i, $X(t)$ is a regenerative process with renewal times being the successive visits to i. The first cycle is τ_i and its distribution is of lattice of span d. By Theorem 3.26 (Smith's Theorem), for $0 \le k < d$,

$$
\begin{aligned}
\lim_{n\to\infty} p_{ij}^{(k+nd)} &= \lim_{n\to\infty} P_i[X(k+nd) = j] \\
&= \frac{d}{E_i(\tau_i)} \sum_{n=0}^{\infty} P_i[X_{k+nd} = j, k+nd < \tau_i]. \quad (4.45)
\end{aligned}
$$

If $d = 1$, then with $k = 0$, (4.45) may be written as

$$
\begin{aligned}
\lim_{n\to\infty} p_{ij}^{(n)} &= \frac{E_i[\#\{n \ge 0;\ X_n = j, n < \tau_i\}]}{E_i(\tau_i)} \\
&= \frac{E_i[\#\{n > 0;\ X_n = j, n \le \tau_i\}]}{E_i(\tau_i)} = \pi_j
\end{aligned}
$$

by (4.41). This proves (4.43). If $d > 1$ and $X_0 = j$, then $X_m = j$ only when m is a multiple of d and hence τ_j is a multiple of d. Letting $k = 0$ and $i = j$ in (4.45) yields

$$\lim_{n\to\infty} p_{jj}^{(nd)} = \frac{dE_j[\#\{n \ge 0;\ X_{nd} = j, nd < \tau_j\}]}{E_j(\tau_j)} = \frac{d}{E_j(\tau_j)} = d\pi_j. \ \lozenge$$

Example 4.36 Consider an MC on three states as in Example 4.23. Its stationary distribution has been determined to be

$$\pi_1 = 0.3571, \qquad \pi_2 = 0.3333, \qquad \pi_3 = 0.3095.$$

From its transition probability matrix $\mathbf{P}$ given in Example 4.23, the MC is clearly aperiodic. By Theorem 4.35,

$$
\lim_{n\to\infty} \mathbf{P}^n = \lim_{n\to\infty}
\begin{bmatrix}
0.2 & 0.3 & 0.5 \\
0.3 & 0.4 & 0.3 \\
0.6 & 0.3 & 0.1
\end{bmatrix}^n
=
\begin{bmatrix}
0.3571 & 0.3333 & 0.3095 \\
0.3571 & 0.3333 & 0.3095 \\
0.3571 & 0.3333 & 0.3095
\end{bmatrix}.
$$

Using MATLAB®, we may compute explicitly to verify the above limit as

$$
\begin{bmatrix} 0.2 & 0.3 & 0.5 \\ 0.3 & 0.4 & 0.3 \\ 0.6 & 0.3 & 0.1 \end{bmatrix}^5 = \begin{bmatrix} 0.3526 & 0.3333 & 0.3141 \\ 0.3567 & 0.3333 & 0.3100 \\ 0.3629 & 0.3333 & 0.3038 \end{bmatrix},
$$

$$
\begin{bmatrix} 0.2 & 0.3 & 0.5 \\ 0.3 & 0.4 & 0.3 \\ 0.6 & 0.3 & 0.1 \end{bmatrix}^{10} = \begin{bmatrix} 0.3572 & 0.3333 & 0.3095 \\ 0.3571 & 0.3333 & 0.3095 \\ 0.3571 & 0.3333 & 0.3096 \end{bmatrix}.
$$

Exercise 4.25 Let X_n be an MC on three states $1, 2, 3$ with the following transition probability matrix,

$$
\mathbf{P} = \begin{bmatrix} 1/2 & 1/2 & 0 \\ 1/3 & 1/3 & 1/3 \\ 0 & 1/2 & 1/2 \end{bmatrix}.
$$

(a) Find $E_1(\tau_1)$ and $E_1(\tau_3)$, the mean return time to state 1 and the mean time to state 3 from state 1.
(b) Find $\lim_{n \to \infty} \mathbf{P}^n$.

Exercise 4.26 Let X_n be an irreducible MC with transition probabilities p_{ij} and period $d > 1$ as in Exercise 4.14. Assume it is positive recurrent with stationary distribution π_j.
(a) Recall that the state space S is divided into d nonempty subsets F_α for $0 \le \alpha < d$, and the restriction of X_n to each F_α and to the time index set $\{0, d, 2d, \ldots\}$ is an irreducible, aperiodic, and positive recurrent MC X_n^* in the state space F_α. Show that the stationary distribution of X_n^* is $\pi_j^* = d\pi_j$ for $j \in F_\alpha$. Consequently, $\lim_{n \to \infty} p_{ij}^{(nd)} = d\pi_j$ for $i, j \in F_\alpha$.
(b) Recall the integer $k(i, j)$ with $0 \le k(i, j) < d$ is defined for any two states i and j to be the unique integer k with $0 \le k < d$ such that $p_{ij}^{(m)} = 0$ for any integer $m > 0$ unless $m = k + nd$ for some integer $n \ge 0$. Show that for any integer k with $0 \le k < d$,

$$
\lim_{n \to \infty} p_{ij}^{(k+nd)} = d\pi_j
$$

if $k = k(i, j)$, and $p_{ij}^{(k+nd)} = 0$ for any integer $n \ge 0$ if $k \ne k(i, j)$.

Exercise 4.27 Let $\mathbf{P}$ be the transition matrix of the symmetric reflecting random walk on $\{0, 1, 2, 3, 4\}$. Based on the results in Exercise 4.26, find $\lim_{n \to \infty} \mathbf{P}^{2n}$ and $\lim_{n \to \infty} \mathbf{P}^{2n+1}$.

Chapter 5

Continuous time Markov chain

5.1 Markov property and transition probability

Definition: A continuous time and discrete state process $X(t)$ is called a continuous time Markov chain if it has the following Markov property in continuous time:

For any $0 \leq s_1 < s_2 < \cdots < s_n < s$, $t > 0$, and states $i_1, i_2, \ldots, i_n, i, j$,

$$P[X(s+t) = j \mid X(s_1) = i_1, X(s_2) = i_2, \ldots, X(s_n) = i_n, X(s) = i]$$
$$= P[X(s+t) = j \mid X(s) = i]. \tag{5.1}$$

This means that given the present (time s), the distribution of the future (time $s+t$) is independent of the past (time $< s$).

Transition probability functions:

$$P_{ij}(t) = P[X(s+t) = j \mid X(s) = i]. \tag{5.2}$$

This is the probability of transition from state i at time s to state j at time $s+t$. The MC (Markov chain) is called time homogeneous if this probability is independent of starting time s. All MCs in the sequel will be time homogeneous unless explicitly stated otherwise.

The Markov property together with time homogeneity may be written as, for $s_1 < s_2 < \cdots < s_n < s$ and $t > 0$,

$$P[X(s+t) = j \mid X(s_1) = i_1, X(s_2) = i_2, \ldots, X(s_n) = i_n, X(s) = i]$$
$$= P_{ij}(t). \tag{5.3}$$

Basic properties of $P_{ij}(t)$:

(a) $P_{ij}(0) = \delta_{ij}$, that is, $P_{ii}(0) = 1$ and $P_{ij}(0) = 0$ for $i \neq j$;

(b) $0 \leq P_{ij}(t) \leq 1$ and $\sum_j P_{ij}(t) = 1$ for any $t \geq 0$; and

(c) (Chapman-Kolmogorov identity): $P_{ij}(t+s) = \sum_k P_{ik}(s)P_{kj}(t)$.

The first two properties (a) and (b) follow directly from the definition of $P_{ij}(t)$, and (c) may be derived using the total probability law as in the case of a discrete time MC (see §4.3).

Transition probability function in matrix form: The transition probability functions together may be regarded as a matrix valued function $\mathbf{P}(t) = \{P_{ij}(t)\}$ with $\mathbf{P}(0) = I$ (the identity matrix). The Chapman-Kolmogorov identity may be written more concisely as

$$\mathbf{P}(s+t) = \mathbf{P}(s)\mathbf{P}(t).$$

Distribution of the Markov chain: By the Markov property, the distribution of the MC $X(t)$ as a process is completely determined by its transition probability function $P_{ij}(t)$ and the initial distribution $p_i = P[X(0) = i]$: For $t_1 < t_2 < \cdots < t_n$,

$$
\begin{aligned}
&P[X(t_1) = i_1, X(t_2) = i_2, \ldots, X(t_n) = i_n] \\
&= \sum_i p_i P_{ii_1}(t_1) P_{i_1 i_2}(t_2 - t_1) \cdots P_{i_{n-1} i_n}(t_n - t_{n-1}). \quad (5.4)
\end{aligned}
$$

Mean occupation time: The amount of time the MC spends in a state j is called the occupation time in j. As before, for any state i, let P_i and E_i be, respectively, the conditional probability and the conditional expectation given $X(0) = i$. The mean occupation time in state j by time t may be written as an integral of transition probability function:

$$
\begin{aligned}
E_i[\text{amount of time MC in } j \text{ by time } t] &= E_i\left[\int_0^t 1_{\{j\}}(X(u))du\right] \\
&= \int_0^t P_{ij}(u)du. \quad (5.5)
\end{aligned}
$$

A more useful form of Markov property: As for a discrete time MC, the Markov property together with time homogeneity may be written in the following more useful and symmetric form: Fix $t > 0$. For $0 \le u_1 < u_2 < \cdots < u_n < t < v_1 < v_2 < \cdots < v_m$, let

$$
\begin{aligned}
A_t &= [X(u_1) = i_1, X(u_2) = i_2, \ldots, X(u_n) = i_n], \\
B_t &= [X(v_1) = j_1, X(v_2) = j_2, \ldots, X(v_m) = j_m], \\
B_0 &= [X(v_1 - t) = j_1, X(v_2 - t) = j_2, \ldots, X(v_m - t) = j_m],
\end{aligned}
$$

where A_t is an event before time t, B_t is an event after time t, and B_0 is the event B_t time shifted backward by t. Then

$$P[B_t \mid A_t, X(t) = i] = P[B_t \mid X(t) = i] = P_i(B_0). \qquad (5.6)$$

In (5.6), A_t and B_t may be any two events determined by the MC before and after time t, respectively.

Regularity assumption: A continuous time MC $X(t)$ is called regular if it has right continuous paths and it can make only finitely many transitions a.s. over any finite time interval. In the sequel, all continuous time MCs are assumed to be regular.

Strong Markov property: As for a discrete time MC, in the Markov property (5.6), the constant time t may be replaced by a stopping time τ of the MC $X(t)$, that is,

$$\begin{aligned} P[B_\tau \mid A_\tau, X(\tau) = i, \tau < \infty] &= P[B_\tau \mid X(\tau) = i, \tau < \infty] \\ &= P_i(B_0), \qquad (5.7) \end{aligned}$$

where A_τ and B_τ are the events determined by the MC before and after τ, respectively, and B_0 is the event B_τ time shifted backward by τ.

The strong Markov property may be proved easily for a discrete stopping time τ. The general case can be derived from a sequence of discrete stopping times $\tau_n \downarrow \tau$ and the right continuity of paths.

5.2 Transition rates

Sojourn time: Let σ be the amount of time the MC $X(t)$ spends in a state before the next transition, called the sojourn time in that state. By the regularity assumption, $X(t)$ has right continuous paths and hence $\sigma > 0$. As before, we will write P_i for the conditional probability given $X(0) = i$.

Theorem 5.1 *For any state i, σ is either exponential or is almost surely ∞ under P_i.*

Proof: Note that $[\sigma > t + s] = [\sigma > t, \sigma_t > s]$, where σ_t is the amount of time between time t and the next transition. Because the event $[\sigma > s]$ is obtained from $[\sigma_t > s]$ after a backward time shift by t, and the event $[\sigma > t]$ may be determined by observing the process up to time t, by the Markov property, we have

$$P_i(\sigma > t + s \mid \sigma > t) = P[\sigma_t > s \mid \sigma > t, X(t) = i] = P_i(\sigma > s).$$

This shows that σ has the lack of memory property, and hence it is exponential unless it is identically ∞. $\Diamond$

Transition rate q_i from a state i: For any state i, let q_i be the exponential rate of the sojourn time σ given $X(0) = i$, that is, $q_i = 1/E_i(\sigma)$, setting $q_i = 0$ if $E_i(\sigma) = \infty$. This may be regarded as the average number of transitions from state i that the MC makes per unit time, called the transition rate from i.

Theorem 5.2 *Assume $P_i(\sigma < \infty) = 1$. Then the sojourn time σ and the shifted process $X^\sigma(t) = X(\sigma + t)$ are independent under P_i.*

Proof: For simplicity, we will only prove the independence of σ and $X^\sigma(t)$ at a fixed $t \geq 0$. The independence of σ and the process at several different time points is proved in the same way, but the notation is more complicated. For any $u > 0$ and state j,

$$P_i[\sigma > u, X^\sigma(t) = j] = \frac{P[\sigma > u, X(0) = i, X(\sigma + t) = j]}{P_i[X(0) = i]}$$

$$= \frac{P[\sigma > u, X(0) = i, X(u) = i, X(u + \sigma_u + t) = j]}{P_i[X(0) = i]}$$

(under P_i, $X(u) = i$ and $\sigma = u + \sigma_u$ if $\sigma > u$)

$$= P[X(u + \sigma_u + t) = j \mid X(0) = i, \sigma > u, X(u) = i]$$
$$\times P[\sigma > u, X(u) = i \mid X(0) = i]$$

$$= P[X(u + \sigma_u + t) = j \mid X(u) = i] P[\sigma > u, X(u) = i \mid X(0) = i]$$

(Markov property at time u)

$$= P[X(\sigma + t) = j \mid X(0) = i] P[\sigma > u \mid X(0) = i] \quad \text{(time shift)}$$

$$= P_i[X^\sigma(t) = j] P_i(\sigma > u). \quad \Diamond$$

Transition probabilities and embedded discrete time MC: For two distinct states i and j, let $p_{ij} = P_i[X(\sigma) = j, \sigma < \infty]$, the probability that the MC makes a transition to state j given it starts in i. Set $p_{ii} = 0$ if $P_i(\sigma < \infty) = 1$ and $p_{ii} = 1$ if $P_i(\sigma = \infty) = 1$.

Let $\sigma_1 = \sigma$, the time of the first transition, and for any integer $n \geq 1$, let σ_n be the time of the nth transition. These are stopping times of the MC $X(t)$. By the strong Markov property of $X(t)$ at times σ_n, it can be shown that $X_n = X(\sigma_n)$ is a discrete time MC with transition probabilities p_{ij}, called the discrete time MC embedded in the continuous time MC $X(t)$. By Theorem 5.2, the embedded discrete time MC is run independently of the sojourn times.

Transition rate q_{ij} from i to j: For two different states i and j, let

$$q_{ij} = q_i p_{ij}. \tag{5.8}$$

This may be regarded as the average number of transitions from i to j per unit time, called the transition rate from i to j. Note that $q_i = \sum_{j \neq i} q_{ij}$.

It is useful to think that from each state i, transitions to other states j occur at independent exponential times of rates q_{ij}, ($q_{ij} = 0$ means no transition from i to j is possible). The transition time from i is the minimum of the transition times to other states and hence has exponential rate $q_i = \sum_{j \neq i} q_{ij}$.

Existence of MC: Given transition rates q_{ij}, that is, given a set of nonnegative numbers q_{ij}, one for each pair of states (i, j), a continuous time MC $X(t)$ may be constructed as follows. Set $q_i = \sum_{j \neq i} q_{ij}$ and $p_{ij} = q_{ij}/q_i$ with $p_{ij} = 0$ when $q_i = 0$. Starting from any state i, let $X(t) = i$ for an exponential time σ of rate q_i ($\sigma = \infty$ when $q_i = 0$); then let a transition to state j take place with probability p_{ij} and let $X(t) = j$ for an exponential time of rate q_j independently of σ. Continue in this way and let $\sigma_1 = \sigma < \sigma_2 < \sigma_3 < \ldots$ be the successive transition times. Suppose $\sigma_n \uparrow \infty$ a.s. Then we obtain a process $X(t)$ for any $t > 0$. The lack of memory of exponential times implies the Markov property and hence $X(t)$ is a continuous time MC with transition rates q_{ij}. By its construction, $X(t)$ is clearly regular.

However, for the above construction to work, we must have $\sigma_n \uparrow \infty$ a.s. We will show this holds under either of the two conditions below:

(a) q_i are bounded, that is, $q_i \leq c$ for all i and some finite constant $c > 0$; or

(b) the discrete time MC with transition probabilities p_{ij} is recurrent.

The proof under (a) is left to Exercise 5.2. Assume (b). Then the number of steps it takes for the discrete time MC to return to i is finite a.s. For the continuous time MC, starting from i, the successive inter-return times τ^n to i are iid of positive mean and each is the sum of finitely many sojourn times. By the SLLN, $\sum_{n=1}^{\infty} \tau^n = \infty$ a.s. This implies $\sigma_n \uparrow \infty$ a.s.

From $P_{ij}(t)$ to q_{ij}: The next theorem provides an expression for the transition rates q_{ij} in terms of the transition probability functions $P_{ij}(t)$. Later in the discussion of Q-matrix, we will obtain an expression for $P_{ij}(t)$ in terms of q_{ij} in the case of finitely many states.

Theorem 5.3 *Let σ be the sojourn time and assume $P_i(\sigma < \infty) = 1$. Then $P_{ij}(t)$ has a continuous derivative in t, and*

$$q_{ij} = \lim_{t \to 0} \frac{1}{t} P_{ij}(t) \text{ for } j \neq i, \quad \text{and} \quad q_i = \lim_{t \to 0} \frac{1}{t}[1 - P_{ii}(t)]. \quad (5.9)$$

Note that if $P_i(\sigma < \infty) = 0$, then $P_{ij}(t) = \delta_{ij}$ for any $t \geq 0$.

Proof: By the total probability law conditioning on σ,

$$
\begin{aligned}
P_{ij}(t) &= e^{-q_i t}\delta_{ij} + \int_0^t \sum_{k \neq j} p_{ik} P_{kj}(t - u) q_i e^{-q_i u} du \\
&= e^{-q_i t}[\delta_{ij} + \int_0^t \sum_{k \neq j} p_{ik} P_{kj}(v) q_i e^{q_i v} dv]. \quad (5.10)
\end{aligned}
$$

Because the integrand of the above integral is bounded, the integral and hence $P_{ij}(t)$ are continuous in t. Then the integrand is continuous and hence $P_{ij}(t)$ is differentiable in t with a continuous derivative. Evaluating the derivative of $P_{ij}(t)$ at $t = 0$ yields (5.9). $\diamond$

Q-matrix (infinitesimal generator): Let $\mathbf{P}(t)$ be the transition probability function in matrix form. The Q-matrix (or infinitesimal generator) is its derivative at $t = 0$, that is,

$$Q = \mathbf{P}'(0) = \lim_{h \to 0} \frac{1}{h}[\mathbf{P}(h) - I]. \quad (5.11)$$

By Theorem 5.3, $Q_{ii} = -q_i$ and $Q_{ij} = q_{ij}$ for $i \neq j$, and thus each row of a Q-matrix sums to 0.

By Chapman-Kolmogorov identity $\mathbf{P}(t + h) = \mathbf{P}(t)\mathbf{P}(h)$,

$$\mathbf{P}'(t) = \mathbf{P}(t)\mathbf{P}'(0) = \mathbf{P}(t)Q. \quad (5.12)$$

The above computation is valid even in the case of infinitely many states because by the mean value theorem in calculus, $(1/h)[P_{ij}(h) - \delta_{ij}] = P'_{ij}(h') \to P'_{ij}(0)$ boundedly as $h \to 0$, where $h' \in (0, h)$.

When there are finitely many states, the matrix form differential equation (5.12) can be solved using the matrix exponential and we obtain

$$\mathbf{P}(t) = e^{Qt} = I + \sum_{n=1}^{\infty} \frac{1}{n!} Q^n t^n. \qquad (5.13)$$

Using the MATLAB® matrix exponential function expm, the above formula may be used to evaluate transition probability $P_{ij}(t)$.

Example 5.4 Consider a continuous time Markov chain with three states $1, 2, 3$ and transition rates $q_{12} = q_{13} = 2$, $q_{21} = 4$, $q_{23} = 3$, $q_{31} = q_{32} = 5$. Find the Q-matrix and the transition probability $P_{12}(0.5)$ from 1 to 2 in time $t = 0.5$.

Solution: The Q-matrix is

$$Q = \begin{bmatrix} -4 & 2 & 2 \\ 4 & -7 & 3 \\ 5 & 5 & -10 \end{bmatrix}.$$

The transition matrix in time $t = 0.5$ is

$$\mathbf{P}(0.5) = e^{0.5Q} = \exp\left\{ 0.5 \begin{bmatrix} -4 & 2 & 2 \\ 4 & -7 & 3 \\ 5 & 5 & -10 \end{bmatrix} \right\}$$

$$= \exp\left\{ \begin{bmatrix} -2 & 1 & 1 \\ 2 & -3.5 & 1.5 \\ 2.5 & 2.5 & -5 \end{bmatrix} \right\} = \begin{bmatrix} 0.5314 & 0.2799 & 0.1887 \\ 0.5142 & 0.2938 & 0.1920 \\ 0.5174 & 0.2896 & 0.1930 \end{bmatrix}$$

(using MATLAB® function expm). Then $P_{12}(0.5) = 0.2799$.

From discrete time to continuous time: Let $\tilde{X}_n$ be a discrete time MC with transition probabilities $\tilde{p}_{ij}$. If we let the process spend an independent exponential time of rate $\tilde{q}_i$ in each state i, instead of one unit time, we obtain a continuous time MC $X(t)$. As a discrete time MC is allowed to make a transition back to the same state, that is, $\tilde{p}_{ii} > 0$ is allowed, the transition rates q_i and transition probabilities p_{ij} of $X(t)$ are not in general equal to $\tilde{q}_i$ and $\tilde{p}_{ij}$. Ignoring the transition back to the same state, we have

$$q_{ij} = \tilde{q}_i \tilde{p}_{ij} \quad \text{for } j \neq i, \qquad (5.14)$$

and hence,

$$q_i = \tilde{q}_i(1 - \tilde{p}_{ii}) \quad \text{and} \quad p_{ij} = \frac{\tilde{p}_{ij}}{1 - \tilde{p}_{ii}} \quad \text{for } i \neq j, \tag{5.15}$$

setting $p_{ij} = 0$ when $\tilde{p}_{ii} = 1$. Note that if $\tilde{p}_{ii} = 0$, then $q_i = \tilde{q}_i$ and $p_{ij} = \tilde{p}_{ij}$. A more formal proof is left to Exercise 5.3.

Exercise 5.1 A salesman travels among three cities 1, 2 and 3. He spends an exponential time in each city, of means 2, 3, and 4 respectively, and then goes to another city. Suppose from city 1 or 2, he goes to one of two other cities with a probability of $1/2$, but from city 3, he goes to 1 or 2 or remains at 3 with a probability of $1/3$. Let $X(t)$ be his location at time t. Then $X(t)$ is a continuous time MC.
(a) Find the Q-matrix.
(b) Find the probability that he is in city 1 at time 10 given he is in city 2 at time 6.

Exercise 5.2 Let σ_n be the successive transition times as defined in the subsection titled "Existence of MC" and assume (a) there. Show that $\sigma_n \uparrow \infty$ a.s. as $n \uparrow \infty$.
Hint: Fix an initial state i and let $z_n = \sigma_n - \sigma_{n-1}$. By the construction, given $X(\sigma_{n-1}) = j$, z_n is exponential of mean $1/q_j$, but if q_j is not a constant, then z_n may not be exponential under P_i. You may try to shorten z_n so that it has the same mean for all j, then z_n become iid exponential under P_i and by SLLN, $\sum_{n=1}^{\infty} z_n = \infty$ a.s.

Exercise 5.3 Prove (5.15).

5.3 Stationary distribution and limiting properties

Stationary distribution: Let $X(t)$ be a continuous time MC with transition probability functions $P_{ij}(t)$. A distribution $\{\pi_j\}$ on its state space is called stationary if for any state j and time t,

$$\pi_j = \sum_i \pi_i P_{ij}(t). \tag{5.16}$$

This means that if the MC $X(t)$ starts with a stationary distribution $\{\pi_j\}$ as the initial distribution, then its distribution at any time $t > 0$

are the same, that is, if $P[X(0) = j] = \pi_j$ for all j, then $P[X(t) = j] = \pi_j$ for all j and $t > 0$. In fact, in this case, the MC is a stationary process as defined in §3.7, that is, for any $0 \le u_1 < u_2 < \cdots < u_n$ and $t > 0$,

$$P[X(u_1 + t) = j_1, X(u_2 + t) = j_2, \ldots, X(u_n + t) = j_n]$$
$$= P[X(u_1) = j_1, X(u_2) = j_2, \ldots, X(u_n) = j_n].$$

This may be derived from (5.16) using the Markov property.

If an MC $X(t)$ starts with a stationary distribution and hence is a stationary process, then it is said to be in the steady state.

Irreducible and ergodic Markov chain: A continuous time MC $X(t)$ is called irreducible, transient, recurrent, positive, or null recurrent if the embedded discrete time MC X_n is so. It is called ergodic if it is irreducible and has a stationary distribution. Note that it is possible that a stationary distribution exists for X_n but not for $X(t)$ (see Theorem 5.6 below), and vice versa (see Exercise 5.7).

Entrance times: Let T_i be the first time when the MC $X(t)$ enters state i from a different state. It is defined to be ∞ if the MC never enters i or never leaves i. Note that T_i is a stopping time of the MC and $T_i > 0$ even when the MC starts in i.

Theorem 5.5 *Let $X(t)$ be an irreducible continuous time MC.*
(a) If $E_i(T_i) < \infty$ for some i, then $X(t)$ is ergodic. Conversely, if $X(t)$ is ergodic, then it is recurrent and $E_i(T_i) < \infty$ for all i.
(b) If $X(t)$ is ergodic, then it has a unique stationary distribution $\{\pi_j\}$ and all $\pi_j > 0$. Moreover, for any two states i and j,

$$\pi_j = \lim_{t \to \infty} P_{ij}(t) \tag{5.17}$$

$$= \lim_{t \to \infty} \frac{time\ in\ state\ j\ by\ time\ t}{t} \quad a.s.\ under\ P_i. \tag{5.18}$$

Thus, the stationary distribution π_j is equal to both the long-run probability that the MC is in state j and the long-run fraction of time it is in j, independently of the starting state i.

Proof: Suppose $E_i(T_i) < \infty$ for some i. For $n \ge 1$, let $T_i^{(n)}$ be the nth time when the MC enters i from a different state and set $T_i^{(0)} = 0$. These are stopping times with $T_i^{(1)} = T_i$. Because T_i is the sum of an exponential sojourn time and an independent random variable, it is nonlattice. By the strong Markov property at $T_i^{(n)}$, $T_i^{(n+1)} - T_i^{(n)}$ are iid of finite mean under P_i and hence $X(t)$ is a regenerative process based on the renewal sequence $T_i^{(n)}$. By Smith's theorem and the long-run average (3.38), the limits in (5.17) and (5.18) are equal. Let π_j denote this common value. Then

$$
\begin{aligned}
\pi_j &= \lim_{u \to \infty} P_{ij}(u) = \lim_{u \to \infty} P_{ij}(u+t) \\
&= \lim_{u \to \infty} \sum_k P_{ik}(u) P_{kj}(t) = \sum_k \pi_k P_{kj}(t).
\end{aligned}
$$

This proves that π_j is a stationary distribution and hence $X(t)$ is ergodic. The irreducibility assumption implies that for any states i and j, $P_{ij}(t) > 0$ for some $t > 0$. Thus, if π_j is a stationary distribution, then $\pi_j = \sum_i \pi_i P_{ij}(t) > 0$. If $\{b_j\}$ is another stationary distribution, then $b_i = \sum_i b_i P_{ij}(t) \to \sum_i b_i \pi_j = \pi_j$ as $t \to \infty$ by (5.17). It remains to prove the second half of (a).

Assume $X(t)$ is ergodic with a stationary distribution $\{\pi_j\}$. Integrating $\pi_j = \sum_i \pi_i P_{ij}(s)$ for s from 0 to t and then dividing by t yields

$$
\pi_j = \sum_i \pi_i \frac{1}{t} \int_0^t P_{ij}(s) ds = \sum_i \pi_i E_i [\frac{\text{time in } j \text{ by time } t}{t}]. \tag{5.19}
$$

If $X(t)$ is transient, then the last expression in (5.19) converges to 0 as $t \to \infty$. This implies $\pi_j = 0$ and so is impossible. Therefore, $X(t)$ is recurrent. Considering a renewal reward process with successive renewals at $T_j^{(n)}$ and a reward being the time in j shows that under P_i,

$$
\lim_{t \to \infty} \frac{\text{time in } j \text{ by time } t}{t} = \frac{E_i[\text{time in } j \text{ in a cycle}]}{E_i(T_j^{(2)} - T_j^{(1)})} = \frac{1/q_j}{E_j(T_j)}
$$

by the Markov property at time T_j. Taking expectation and then substituting in (5.19), we obtain $\pi_j = (1/q_j)/E_j(T_j)$, and because $\pi_j > 0$, $E_j(T_j) < \infty$ for any j. $\diamond$

Theorem 5.6 *Let $X(t)$ be an irreducible and positive recurrent continuous time MC, and let $\{\eta_j\}$ be the stationary distribution of the embedded discrete time MC X_n (the existence and uniqueness of $\{\eta_j\}$ is guaranteed by Theorem 4.34). Then $X(t)$ is ergodic if and only if*

$\sum_i \eta_i/q_i < \infty$. *Moreover, in the ergodic case, the stationary distribution* $\{\pi_j\}$ *of* $X(t)$ *is given by*

$$\pi_j = \frac{\eta_j/q_j}{\sum_i \eta_i/q_i}. \tag{5.20}$$

Proof: Considering the MC $X(t)$ as a regenerative process under P_i based on successive returns to state i as in the proof of Theorem 5.5. By Corollary 4.29, η_j/η_i is the expected number of visits to j between two successive visits to i. Because the mean time spent in j for each visit is $1/q_j$, it follows that the mean time in j in a cycle is $(\eta_j/q_j)/\eta_i$ and the mean cycle length of the regenerative process is $(\sum_j \eta_j/q_j)/\eta_i$. By Theorem 5.5, $X(t)$ is ergodic if and only if $(\sum_j \eta_j/q_j)/\eta_i < \infty$. In this case, let

$$\pi_j = \frac{\eta_j/q_j}{\sum_k \eta_k/q_k} = \frac{(\eta_j/q_j)/\eta_i}{\sum_k (\eta_k/q_k)/\eta_i}.$$

Then $\{\pi_j\}$ is the long-run fraction of time when $X(t)$ is in j. By Smith's Theorem,

$$\pi_j = \lim_{u \to \infty} P_{ij}(t+u) = \lim_{u \to \infty} \sum_k P_{ik}(u)P_{kj}(t) = \sum_k \pi_k P_{kj}(t).$$

This shows that $\{\pi_j\}$ is a stationary distribution. $\diamond$

Balance equations: The stationary distribution $\{\pi_j\}$ is defined by (5.16), and it is determined by this equation together with the normalizing equation $\sum_j \pi_j = 1$, but (5.16) is difficult to solve as the transition probability function $P_{ij}(t)$ is often not explicitly available or too complicated to use. However, the stationary distribution $\{\pi_j\}$ may be conveniently obtained by solving the following balance equations,

$$\pi_j q_j = \sum_{i \neq j} \pi_i q_{ij} \quad \text{(mass out from } j = \text{mass into } j\text{)} \tag{5.21}$$

for all j, together with $\sum_j \pi_j = 1$.

Note that (5.21) is just the derivative of (5.16) at $t = 0$ and has the following simple interpretation: Think of a stationary distribution $\{\pi_j\}$ as unit mass distributed among all states, then the left-hand side of (5.21) is the rate at which mass is moving out of state j and the right-hand side is the rate at which mass is moving into j.

Equivalence: The system of balance equations in (5.21) is equivalent to (5.16). Because (5.21) is just $\sum_i \pi_i P'_{ij}(0) = 0$, if π_j satisfy (5.16), then they also satisfy (5.21). Conversely, if π_j satisfy (5.21), then

$$\frac{d}{dt} \sum_i \pi_i P_{ij}(t) = \frac{d}{du} \sum_i \pi_i P_{ij}(u+t) \mid_{u=0}$$

$$= \frac{d}{du} \sum_{i,k} \pi_i P_{ik}(u) P_{kj}(t) \mid_{u=0} = \sum_{i,k} \pi_i P'_{ik}(0) P_{kj}(t) = 0.$$

This implies that $\sum_i \pi_i P_{ij}(t)$ is a constant and hence is equal to $\sum_i \pi_i P_{ij}(0) = \pi_j$. This is (5.16).

Solving balance equations: The stationary distribution $\{\pi_j\}$ may be determined by solving the balance equations (5.21) together with $\sum_j \pi_j = 1$. Note that the balance equations in (5.21) are redundant because they sum to an identity, and hence one of them may be dropped when solving for the stationary distribution.

For an ergodic MC with finitely many states, the system of balance equations in (5.21) together with $\sum_j \pi_j = 1$ has a unique solution, which is necessarily the stationary distribution. This may be proved in the same way as for a discrete time MC in §4.7.

Example 5.7 Consider a continuous time Markov chain $X(t)$ on three states: 1, 2, and 3. The process spends exponential random times in these states of means 5, 4, and 3, respectively, and then goes to one of the other two states with equal chance.
(a) Find the transition rates q_{ij}.
(b) Find the long-run probabilities that $X(t)$ is in these states.
(c) Evaluate the transition probability function $\mathbf{P}(t) = \{P_{ij}(t)\}$ for $t = 10, 20, 30$ and compare it with the limiting probabilities in (b).
(d) Find the long-run fraction of time the process spends in each state.
(e) Suppose running the process incurs a cost, which is zero in state 1, but 120 and 100 per unit time in states 2 and 3, respectively. Find the long-run average cost per unit time.

Solution: (a) From the mean exponential times, $q_1 = 1/5$, $q_2 = 1/4$, and $q_3 = 1/3$. Since $p_{ij} = 1/2$ and $q_{ij} = q_i p_{ij}$ for $i \neq j$, $q_{12} = q_{13} = 1/10$, $q_{21} = q_{23} = 1/8$, and $q_{31} = q_{32} = 1/6$. These transition rates are depicted in the following transition diagram, which is useful for writing down balance equations.

(1) $1/10 \longrightarrow$ (2)
                       $\longleftarrow 1/8$

       $1/10 \searrow\nwarrow 1/6$ $1/6 \nearrow\swarrow 1/8$

                     (3)

(b) The first two balance equations together with $\sum_j \pi_j = 1$ are

$$\frac{1}{5}\pi_1 = \frac{1}{8}\pi_2 + \frac{1}{6}\pi_3, \quad \frac{1}{4}\pi_2 = \frac{1}{10}\pi_1 + \frac{1}{6}\pi_3, \quad \text{and} \quad \pi_1 + \pi_2 + \pi_3 = 1.$$

Solve them to obtain $\pi_1 = 5/12 = 0.4167$, $\pi_2 = 1/3 = 0.3333$, and $\pi_3 = 1/4 = 0.25$. These are the long-run probabilities that $X(t)$ is in states $1, 2, 3$.

Alternatively, we may use (5.20) to find π_j from the stationary distribution $\{\eta_j\}$ of the embedded discrete time Markov chain. Since the transition from any state to the other two states have equal chance, it is easy to see that $\eta_1 = \eta_2 = \eta_3 = 1/3$. Then

$\pi_1 = (1/3)(5)/[(1/3)(5) + (1/3)(4) + (1/3)(3)] = 5/12 = 0.4167$,
$\pi_2 = (1/3)(4)/4 = 1/3$ and $\pi_3 = (1/3)(3)/4 = 1/4$.

(c) From (a), the Q-matrix is

$$Q = \begin{bmatrix} -1/5 & 1/10 & 1/10 \\ 1/8 & -1/4 & 1/8 \\ 1/6 & 1/6 & -1/3 \end{bmatrix}.$$

Using MATLAB® matrix exponential function expm, we find

$$\mathbf{P}(10) = \exp(10Q) = \begin{bmatrix} 0.4363 & 0.3191 & 0.2446 \\ 0.3989 & 0.3513 & 0.2499 \\ 0.4077 & 0.3331 & 0.2592 \end{bmatrix},$$

$$\mathbf{P}(20) = \exp(20Q) = \begin{bmatrix} 0.4174 & 0.3328 & 0.2498 \\ 0.4160 & 0.3339 & 0.2501 \\ 0.4164 & 0.3335 & 0.2501 \end{bmatrix},$$

$$\mathbf{P}(30) = \exp(30Q) = \begin{bmatrix} 0.4167 & 0.3333 & 0.2500 \\ 0.4166 & 0.3334 & 0.2500 \\ 0.4167 & 0.3333 & 0.2500 \end{bmatrix}.$$

We see that the transition probabilities indeed converge to the stationary distribution.

(d) The long-run fractions of time the process spends in three states are the same as the long-run probabilities found in (b).

(e) The long-run average cost per unit time is

$$0\pi_1 + 120\pi_2 + 100\pi_3 = 0(5/12) + 120(1/3) + 100(1/4) = 65.$$

Uniformization By Theorem 5.6 and Exercise 5.7, the stationary distribution $\{\pi_j\}$ of a continuous time MC $X(t)$ and the stationary distribution $\{\eta_j\}$ of the embedded MC may not exist together, and when they do, they are in general not equal. However, by (5.20), they exist together and are equal if all q_i are the same.

For a general irreducible continuous time MC $X(t)$, assuming its q_i are bounded, say $q_i \leq b$ for some constant $b > 0$, by adding fictitious transitions back to same states, we may view the MC as having a transition rate $\tilde{q}_{ii} = b - q_i$ from i back to i. Then the total transition rate from any i is the constant b, the transition rate from i to $j \neq i$ is still q_{ij} but the transition probability from i to $j \neq i$ is now $\tilde{p}_{ij} = q_{ij}/b$. This is still the same continuous time MC, with the same stationary distribution if one exists, but the embedded discrete time MC has been changed. This procedure for obtaining a constant transition rate is called uniformization.

Example 5.8 To perform uniformization for the continuous time MC $X(t)$ in Example 5.7, we may take $b = 1/3$. Recall $q_{1j} = 1/10$ for $j \neq 1$, $q_{2j} = 1/8$ for $j \neq 2$, and $q_{3j} = 1/6$ for $j \neq 3$. Then

$$\tilde{p}_{1j} = \frac{1/10}{1/3} = \frac{3}{10} \text{ for } j \neq 1, \quad \tilde{p}_{11} = 1 - 2(\frac{3}{10}) = \frac{2}{5},$$

$$\tilde{p}_{2j} = \frac{1/8}{1/3} = \frac{3}{8} \text{ for } j \neq 2, \quad \tilde{p}_{22} = 1 - 2(\frac{3}{8}) = \frac{1}{4},$$

$$\tilde{p}_{3j} = \frac{1/6}{1/3} = \frac{1}{2} \text{ for } j \neq 3, \quad \tilde{p}_{33} = 1 - 2(\frac{1}{2}) = 0.$$

Exercise 5.4 A rat is released into one of six rooms numbered 1 through 6 as shown below. Suppose the rat spends an exponential random time of mean 2 in each room and then moves into an adjacent room chosen at random. Let $X(t)$ be the room number at time t. This is a continuous time Markov chain because the inter-transition times are independent and exponential.

1	2	3
4	5	6

(a) In the long run, what is the probability of finding the rat in each room?

(b) Find the transition probability matrix function $\mathbf{P}(t)$ for $t = 10, 20, 30, 40, 50$ to see how the transition probabilities converge to the stationary distribution found in (a).

Hint: Use symmetry to simplify the computation.

Exercise 5.5 A machine can be in one of three states: breakdown (0), poor working (1), and good working (2). At the good working state, it takes an exponential random time of mean 2 hours before the machine either breaks down or enters the poor working state with equal probability. At the poor working state, it lasts an exponential random time of mean 0.5 hour, then it either breaks down or returns to good working state, but four out five times it breaks down. When it breaks down, it is repaired to good working condition and the repair time is exponential of mean 1 hour.

(a) In the long run, what fraction of time is the machine in good working condition?

(b) Suppose the output of the machine is 20 units per hour at good working condition and 10 units per hour at poor working. What is the long-run average output per hour?

Exercise 5.6 Passengers arrive at a train station according to a Poisson process at a rate of 10 per hour and trains depart the station according to an independent Poisson process at a rate of 1 per hour. Each train carries all passengers presently at the station. Find the long-run fraction of time when there are at least 10 passengers in the station.

Exercise 5.7 The symmetric random walk on integers is known to be irreducible and null recurrent (see Example 4.15). Therefore, it has no stationary distribution as a discrete time MC. Now impose a sojourn time at state j of exponential rate $2^{|j|}$ for all $j = 0, \pm 1, \pm 2, \ldots$. Show that the resulting continuous time MC has a stationary distribution and hence is ergodic.

5.4 Birth and death processes

Basic definition: A continuous time MC $X(t)$ on nonnegative integer states $0, 1, 2, \ldots$ is called a birth and death process, or a BD process, if during each transition, it can only either increase or decrease by 1, that is, if $q_{ij} = 0$ if $i - j \neq \pm 1$. The state of this Markov chain may be regarded as the size of a population.

Birth and death rates: $\lambda_i = q_{i\,i+1}$ is called the birth rate and $\mu_i = q_{i\,i-1}$ is called the death rate at state i. The transition diagram of a BD process is depicted below:

$$(0) \overset{\lambda_0}{\to} \overset{\mu_1}{\leftarrow} (1) \overset{\lambda_1}{\to} \overset{\mu_2}{\leftarrow} (2) \overset{\lambda_2}{\to} \cdots \overset{\mu_n}{\leftarrow} (n) \overset{\lambda_n}{\to} \overset{\mu_{n+1}}{\leftarrow} (n+1) \cdots .$$

Note that $\mu_0 = 0$ (no death can occur when the size of the population is zero), $q_0 = \lambda_0$, $q_i = \lambda_i + \mu_i$ for $i \geq 1$. The process will be called a pure birth process if all $\mu_i = 0$ and a pure death process if all $\lambda_i = 0$.

A Poisson process of rate λ is a pure birth process with a constant birth rate λ. Its transition probability function is clearly

$$P_{ij}(t) = e^{-\lambda t}(\lambda t)^{j-i}/(j-i)! \quad \text{for } i \leq j.$$

Yule process: A Yule process is a pure birth process with birth rate $\lambda_i = \lambda i$ for some constant $\lambda > 0$. In this case, $q_i = q_{i\,i+1} = i\lambda$ and all other $q_{ij} = 0$. Although there are infinitely many states, the matrix differential equation (5.12), $\mathbf{P}'(t) = \mathbf{P}(t)Q$, still makes sense as component-wise it is just

$$P'_{ij}(t) = (j-1)\lambda P_{i\,j-1}(t) - j\lambda P_{ij}(t) \tag{5.22}$$

for all $i \geq 0$ and $j \geq 0$, setting $P_{ij}(t) = 0$ for $j = -1$. Using the initial conditions $P_{ij}(0) = \delta_{ij}$, these differential equations can be solved to obtain

$$P_{ij}(t) = \binom{j-1}{i-1} e^{-\lambda i t}(1 - e^{-\lambda t})^{j-i}. \quad \text{for } j \geq i \geq 1. \tag{5.23}$$

One may verify directly that $P_{ij}(t)$ in (5.23) indeed satisfies the differential equation (5.22) (setting $P_{ij}(t) = 0$ for $j < i$). In particular,

$$P_{1j}(t) = e^{-\lambda t}(1 - e^{-\lambda t})^{j-1} \quad \text{for } j \geq 1. \tag{5.24}$$

From this formula, one sees that if the population (Yule process) starts with 1 individual at time 0, then the size of the population at time t has a geometric distribution with parameter $p = e^{-\lambda t}$. Therefore, the mean size of the population at time t, $E[X(t)]$, is equal to $1/p = e^{\lambda t}$, which grows exponentially in t.

Stationary distribution of a BD process: Assume $\lambda_i > 0$ for all $i \geq 0$ and $\mu_i > 0$ for all $i \geq 1$. Then the BD process is an irreducible MC. Let $\theta_0 = 1$ and $\theta_j = (\lambda_0\lambda_1\cdots\lambda_{j-1})/(\mu_1\mu_2\cdots\mu_j)$ for $j \geq 1$. The balance equations (5.21) together with $\sum_j \pi_j = 1$ can be easily solved to obtain the following stationary distribution:

$$\pi_0 = \frac{1}{\sum_{i=0}^{\infty}\theta_i}, \quad \pi_j = \theta_j\pi_0 = \frac{\theta_j}{\sum_{i=0}^{\infty}\theta_i} \quad \text{for } j = 1, 2, \ldots, \quad (5.25)$$

provided $\sum_{i=0}^{\infty}\theta_i$ converges. If not, there is no stationary distribution.

In the case of constant birth and death rates, that is, when $\lambda_i = \lambda$ for all $i \geq 0$ and $\mu_i = \mu$ for all $i > 0$, there is a stationary distribution if and only if $\lambda < \mu$, and then

$$\pi_j = (\frac{\lambda}{\mu})^j(1 - \frac{\lambda}{\mu}) \quad \text{for } j = 0, 1, 2, \ldots. \quad (5.26)$$

Linear growth with immigration: Consider a population in which each individual may give birth to a new individual at an exponential rate $\lambda > 0$ and may die at an exponential rate $\mu > 0$. The two events occur independently of each other and of other individuals. There is also an independent immigration that causes a unit increase in the size of the population at an exponential rate $\gamma > 0$. Let $X(t)$ be the size of population at time t. Then $X(t)$ is a BD process with birth rate $\lambda_i = \lambda i + \gamma$ and death rate $\mu_i = \mu i$, such a process is called a linear growth model with immigration.

To determine the stationary distribution $\{\pi_j\}$, note that $\theta_0 = 1$ and

$$\begin{aligned}
\theta_j &= \frac{\gamma(\gamma + \lambda)\cdots(\gamma + (j-1)\lambda)}{\mu(2\mu)\cdots(j\mu)} \\
&= \frac{(\gamma/\lambda)(\gamma/\lambda + 1)\cdots(\gamma/\lambda + j - 1)}{j!}(\frac{\lambda}{\mu})^j \\
&= \binom{\gamma/\lambda + j - 1}{j}(\frac{\lambda}{\mu})^j,
\end{aligned}$$

where the well-known notation

$$\binom{n}{j} = \frac{(n - j + 1)(n - j + 2)\cdots(n - 1)n}{j!}$$

for positive integers $n \geq j$ is extended to any real number n. Using the ratio test, it is easy to show that the series $\sum_{i=0}^{\infty} \theta_i$ converges if $\lambda < \mu$ and diverges if $\lambda > \mu$. If $\lambda = \mu$, then $\theta_j \geq \gamma/(j\lambda)$ and hence the series $\sum_{i=0}^{\infty} \theta_i$ diverges. Therefore, there is a stationary distribution $\{\pi_j\}$ if and only if $\lambda < \mu$. Moreover, by the Taylor series $(1-x)^{-n} = \sum_{j=0}^{\infty} \binom{n+j-1}{j} x^j$ for $|x| < 1$, $\sum_{j=0}^{\infty} \theta_j = (1 - \lambda/\mu)^{-\gamma/\lambda}$ provided $\lambda < \mu$. The following result now follows easily.

Theorem 5.9 *Let $X(t)$ be the population size in a linear growth model with birth rate $\lambda > 0$, death rate $\mu > 0$, and immigration rate $\gamma \geq 0$. Then $X(t)$ is an irreducible MC. It is ergodic if and only if $\lambda < \mu$. Moreover, if $\lambda < \mu$, then the unique stationary distribution is given by*

$$\pi_0 = (1 - \frac{\lambda}{\mu})^{\gamma/\lambda}, \quad \pi_j = (\frac{\lambda}{\mu})^j \binom{\gamma/\lambda+j-1}{j}(1 - \frac{\lambda}{\mu})^{\gamma/\lambda} \text{ for } j \geq 1. \quad (5.27)$$

Example 5.10 For a linear growth model with $\lambda = \gamma = 1$ and $\mu = 2$,

$$\pi_0 = (1 - \frac{1}{2})^{1/1} = \frac{1}{2}, \quad \pi_j = (\frac{1}{2})^j \binom{j}{j}(1 - \frac{1}{2})^{1/1} = (\frac{1}{2})^{j+1} \text{ for } j \geq 1.$$

Stationary distribution for a limited population size: If a BD process has zero birth rate when the size of the population reaches a certain limit $N > 0$, that is, if $\lambda_N = q_{N\,N+1} = 0$, then we obtain a continuous time Markov chain on states $0, 1, 2, \ldots, N$, called a BD process with the maximum size N. Its transition diagram is depicted below:

$$(0) \overset{\lambda_0}{\underset{\mu_1}{\rightleftarrows}} (1) \overset{\lambda_1}{\underset{\mu_2}{\rightleftarrows}} (2) \overset{\lambda_2}{\rightarrow} \cdots \overset{\mu_{N-1}}{\leftarrow} (N-1) \overset{\lambda_{N-1}}{\underset{\mu_N}{\rightleftarrows}} (N).$$

If $\lambda_i > 0$ for $0 \leq i \leq N-1$ and $\mu_i > 0$ for $1 \leq i \leq N$, then the BD process is an irreducible MC with finitely many states. It always has a unique stationary distribution $\{\pi_j\}$ given by

$$\pi_0 = \frac{1}{\sum_{i=0}^{N} \theta_i}, \quad \pi_j = \theta_j \pi_0 = \frac{\theta_j}{\sum_{i=0}^{N} \theta_i} \text{ for } 1 \leq j \leq N. \quad (5.28)$$

In the case of constant birth and death rates, that is, if $\lambda_i = \lambda > 0$ for $0 \leq i \leq N-1$ and $\mu_i = \mu > 0$ for $1 \leq i \leq N$, then $\theta_j = (\frac{\lambda}{\mu})^j$ and

$$\pi_j = \frac{(\lambda/\mu)^j(1 - \lambda/\mu)}{1 - (\lambda/\mu)^{N+1}} \text{ for } j = 0, 1, 2, \ldots, N. \quad (5.29)$$

Example 5.11 A factory has three machines and one repairman. A machine is up (working) for an exponential time of mean 5 and then it is down, independently of other machines. A down machine is immediately repaired if the repairman is free, otherwise, its repair is put on hold until the repairman is free. Suppose the repair time is exponential of mean 2. In the long run, what fraction of time are all three machines up and what fraction of time are all three down?

Solution: The number of up machines is a BD process with maximum size $N = 3$, birth rate $\lambda_j = 1/2$ for $j = 0, 1, 2$, and death rate $\mu_j = j/5$ for $j = 1, 2, 3$, noting that μ_j is the rate of the minimum of j exponential random variables of mean 5. Then $\theta_j = (5/2)^j / j!$ for $j = 1, 2, 3$ and $\sum_{j=0}^{3} \theta_0 = 1 + 5/2 + (5/2)^2/2 + (5/2)^3/6 = 9.2292$.

The long-run fraction of time when all three machines are up is $\pi_3 = \theta_3/9.2292 = 0.2822$, and that when all three machines are down is $\pi_0 = 1/9.2292 = 0.1084$.

Exercise 5.8 For the linear growth model in Example 5.10 with $\lambda = \gamma = 1$ and $\mu = 2$, if the population size is capped at $N = 5$, find its stationary distribution.

Exercise 5.9 Solve the problem in Example 5.11 but assume there are two repairmen with exponential repair times of same mean 2 so that two machines may be repaired together.

Exercise 5.10 Solve the problem in Exercise 5.9 but now assume the two repairmen have different mean exponential repair times 2 and 3, and when both repairmen are free, a down machine will be repaired by the fast repairman (mean 2).
Hint: A BD process may not apply here. Note that when there is only 1 machine under repair, it may be repaired by either the fast repairman or the slow one, which corresponds to two different states of the system. Draw a transition diagram.

5.5 Exponential queuing systems

Exponential queuing systems: Queuing systems were introduced in §3.3. We will now assume that the inter-arrival times of customers and service times are independent and exponential. Because of the lack of memory property of the exponential distribution, the state of the system is determined by the total queue length, that is, the number of customers in system, except when the servers are heterogeneous, in which case it is also necessary to specify which servers are currently engaged when not all of them are busy. In any case, because the change of state is exponential, the system is described by an irreducible continuous time MC whose states are the total queue length possibly plus some additional information on how the servers are engaged.

Recall that the system is stable if the mean return time to state 0 is finite. By part (a) of Theorem 5.5, the stability is equivalent to the ergodicity of the MC. Thus, the stability of the queuing system is the same as the existence of a stationary distribution π_j. If the system is stable, then in the long run, both the probability and the fraction of time when the system is in state j are equal to π_j, and the system is said to be in the steady state.

M/M/1: Customers arrive at a single-server station according to a Poisson process of rate λ, and the service times of the single server are iid exponential of rate μ (thus the mean service time is $1/\mu$).

The total queue length process $X(t)$ is a BD process with constant birth rate λ and death rate μ. By (5.26), if $\lambda < \mu$, then there is a unique stationary distribution $\{\pi_j\}$ given by

$$\pi_j = (\lambda/\mu)^j (1 - \lambda/\mu) \quad \text{for } j = 0, 1, 2, \ldots. \tag{5.30}$$

If $\lambda \geq \mu$, then the system is not stable and has no stationary distribution.

M/M/k: Consider a queuing system as above but now assume it has k servers of same exponential service rate μ so that up to k customers may be served at the same time. The total queue length process $X(t)$ is still a BD process with constant birth rate λ and death rate

$$\mu_j = \begin{cases} j\mu, & \text{if } j \leq k \\ k\mu, & \text{if } j > k. \end{cases}$$

The stationary distribution may be obtained from (5.25) with $\theta_j = \lambda^j/(j!\mu^j)$ for $j \leq k$ and $\theta_j = \lambda^j/(k!k^{j-k}\mu^j)$ for $j > k$. It can be shown that the series $\sum_j \theta_j$ converges if and only if $\lambda < k\mu$. Therefore, the stationary distribution exists if and only if $\lambda < k\mu$ (arrival rate is less than total service rate), and it is given by

$$\pi_0 = [\sum_{j=0}^{k-1} \frac{1}{j!}\xi^j + \frac{1}{k!}\xi^k \frac{1}{1-\rho}]^{-1}, \quad \xi = \frac{\lambda}{\mu} \text{ and } \rho = \frac{\lambda}{k\mu},$$

$$\pi_j = \begin{cases} \frac{1}{j!}\xi^j \pi_0 & \text{for } 1 \le j \le k \\ \frac{1}{k!}\xi^k \rho^{j-k} \pi_0 & \text{for } j > k. \end{cases} \tag{5.31}$$

M/M/k/N: This is just like an M/M/k system, but it can have at most N ($\ge k$) customers, including those at service. Arrivals will not enter the system when there are N in the system. The total queue length is a BD process with maximum size N, called the holding capacity. Because of finitely many states, there is always a unique stationary distribution still given by (5.31) except now $j \le N$ and

$$\pi_0 = [\sum_{j=0}^{k-1} \frac{1}{j!}\xi^j + \frac{1}{k!}\xi^k (\frac{1-\rho^{N-k+1}}{1-\rho})]^{-1}. \tag{5.32}$$

In particular, for $j = k = N$, we obtain Erlang's loss formula:

$$\pi_N = \frac{\xi^N/N!}{\sum_{i=0}^{N} \xi^i/i!}, \tag{5.33}$$

which is the long-run fraction of time when customers cannot enter the system because all servers are busy, and are thus lost to the system.

Example 5.12 For a system with exponential arrival rate $\lambda = 3$, $k = 2$ servers of exponential service rate $\mu = 1$ each, and holding capacity $N = 5$, the long-run fraction of time when the system is empty is

$$\pi_0 = [1 + \frac{3}{1} + \frac{1}{2!}(\frac{3}{1})^2(\frac{1-[3/(2\cdot 1)]^{5-2+1}}{1-3/(2\cdot 1)})]^{-1} = 0.0247,$$

and that when the system is full is $\pi_5 = (1/2!)(3/1)^2(3/2)^{5-2}\pi_0 = 0.3751$.

Two heterogeneous servers: Consider Poisson arrivals of rate λ to a system of two heterogeneous servers A and B with exponential service rates μ_A and μ_B, respectively. Assume a customer arriving at an empty system is to be served by server A with probability p and by server B with probability $q = 1 - p$ for $0 \le p \le 1$. The state of the system is determined by the number of customers in the system except when there is 1 customer, then the engaged server is to be indicated. The states are thus $0, 1A, 1B, 2, 3, \ldots$ and transitions are depicted in the following diagram, where $\mu = \mu_A + \mu_B$ is the total service rate.

$$
\begin{array}{c}
(1A) \\
(0) \begin{array}{c} p\lambda \nearrow\hspace{-0.5em}\searrow \mu_A \\ q\lambda \searrow\hspace{-0.5em}\nearrow \mu_B \end{array} \quad \begin{array}{c} \lambda \searrow\hspace{-0.5em}\nearrow \mu_B \\ \lambda \nearrow\hspace{-0.5em}\searrow \mu_A \end{array} (2) \begin{array}{c} \lambda \longrightarrow \\ \longleftarrow \mu \end{array} (3) \begin{array}{c} \lambda \longrightarrow \\ \longleftarrow \mu \end{array} (4) \ \cdots \\
(1B)
\end{array}
$$

The balance equations at states $0, 1A, 1B, 2$, and j for $j \ge 3$ are as follows:

$$
\begin{aligned}
\lambda\pi_0 &= \mu_A\pi_{1A} + \mu_B\pi_{1B} \\
(\lambda + \mu_A)\pi_{1A} &= p\lambda\pi_0 + \mu_B\pi_2 \\
(\lambda + \mu_B)\pi_{1B} &= q\lambda\pi_0 + \mu_A\pi_2 \\
(\lambda + \mu)\pi_2 &= \lambda\pi_{1A} + \lambda\pi_{1B} + \mu\pi_3 \\
(\lambda + \mu)\pi_j &= \lambda\pi_{j-1} + \mu\pi_{j+1} \quad \text{for } j \ge 3.
\end{aligned}
$$

Suppose there is a stationary distribution. Let $\pi_1 = \pi_{1A} + \pi_{1B}$ and let $\nu \le \mu$ be defined by $\nu\pi_1 = \mu_A\pi_{1A} + \mu_B\pi_{1B}$. By the balance equations at $0, 1A, 1B$, and 2,

$$
\lambda\pi_0 = \nu\pi_1, \quad (\lambda + \nu)\pi_1 = \lambda\pi_0 + \mu\pi_2, \quad (\lambda + \mu)\pi_2 = \lambda\pi_1 + \mu\pi_3.
$$

These three equations together with the balance equation at $j \ge 3$ form the system of balance equations for a BD process with a constant birth rate λ, and death rates $\mu_1 = \nu$ and $\mu_j = \mu$ for $j \ge 2$, which represents the original system when the two states $1A$ and $1B$ are combined into a single state 1. This implies that $\sum_j (\lambda/\mu)^j < \infty$, that is, $\rho = \lambda/\mu < 1$, and by (5.25), the stationary distribution of the BD process is given by

$$
\pi_0 = [1 + \frac{\lambda/\nu}{1 - \rho}]^{-1}, \quad \pi_j = (\frac{\lambda}{\nu})\rho^{j-1}\pi_0 \quad \text{for } j \ge 1. \tag{5.34}
$$

Now π_{1A} and π_{1B} may be solved from the balance equations at $1A$ and $1B$ in terms of ν, and then from $\mu_{1A} + \mu_{1B} = \pi_1$, we may determine ν. The results are summarized as follows.

Theorem 5.13 *For a queuing system with two heterogeneous servers*

as described above, the stationary distribution $\{\pi_0, \pi_{1A}, \pi_{1B}, \pi_2, \pi_3, \ldots\}$
exists if and only if $\rho < 1$, *where* $\rho = \lambda/(\mu_A+\mu_B)$ *is the traffic intensity,
and is given by (5.34) together with*

$$\pi_{1A} = [\frac{p\lambda}{\lambda + \mu_A} + \frac{\mu_B}{\lambda + \mu_A}(\frac{\lambda\rho}{\nu})]\pi_0, \qquad (5.35)$$

$$\pi_{1B} = [\frac{q\lambda}{\lambda + \mu_B} + \frac{\mu_A}{\lambda + \mu_B}(\frac{\lambda\rho}{\nu})]\pi_0, \qquad (5.36)$$

where

$$\nu = \frac{(2\rho + 1)\mu_A\mu_B}{\lambda + p\mu_B + q\mu_A}. \qquad (5.37)$$

Moreover, π_1 *in (5.34) in equal to* $\pi_{1A} + \pi_{1B}$,

$$\lambda\pi_0 = \nu\pi_1 = \mu_A\pi_{1A} + \mu_B\pi_{1B}, \qquad (5.38)$$

and $\{\pi_0, \pi_1, \pi_2, \ldots\}$ *is the stationary distribution of a BD process with
a constant birth rate* λ, *and death rates* $\mu_1 = \nu$ *and* $\mu_j = \mu$ *for* $j \geq 2$.

Two heterogeneous servers with a finite holding capacity: In
the above discussion, if the queuing system has a finite holding capacity
$N \geq 2$, then the stationary distribution always exists and is still given
by the formulas in Theorem 5.13 except now $j \leq N$ and

$$\pi_0 = [1 + (\frac{\lambda}{\nu})\frac{1 - \rho^N}{1 - \rho}]^{-1}. \qquad (5.39)$$

Two heterogeneous servers with $p = 1/2$**:** In this case, a simple
computation of (5.37) yields $\nu = 2\mu_A\mu_B/(\mu_A + \mu_B)$, and

$$\pi_{1A} = \frac{1}{2}\frac{\lambda}{\mu_A}\pi_0, \qquad \pi_{1B} = \frac{1}{2}\frac{\lambda}{\mu_B}\pi_0, \qquad \pi_j = \frac{1}{2}\frac{\lambda^2}{\mu_A\mu_B}\rho^{j-2}\pi_0$$

for $j \geq 2$. This holds for both infinite and finite holding capacity.

Multiple heterogeneous servers: In a queuing system with Poisson
arrival rate λ and k exponential servers with rates $\mu_1, \ldots, \mu_k$, assuming
no waiting space and when there are more than one available servers,
they are equally likely to be chosen by an arriving customer, it can be
shown that the stationary distribution

$$\tilde{\pi}(x_1, \ldots, x_k) = P[X_1 = x_1, \ldots, X_k = x_k], \qquad x_i = 0 \text{ or } 1,$$

where $X_i = 0$ or 1 means server i is idle or busy, is given by

$$\tilde{\pi}(x_1,\ldots,x_k) = \binom{k}{x_1+\cdots+x_k}^{-1}\tilde{\pi}_0\frac{(\lambda/\mu_1)^{x_1}\cdots(\lambda/\mu_k)^{x_k}}{(x_1+\cdots+x_k)!}, \qquad (5.40)$$

where $\tilde{\pi}_0 > 0$ is a normalizing constant.

In the case of unlimited waiting space, if $\rho = \lambda/(\mu_1 + \cdots + \mu_k)$ is less than 1, then the stationary distribution

$$\pi(x_1,\ldots,x_k;m) = P[X_1 = x_1,\ldots,X_k = x_k, Q = m]$$

for $x_i = 0$ or 1, $m = 0,1,2,3,\ldots$, where Q is the number of customers waiting in queue, exists and is given by

$$\begin{aligned} \pi(x_1,\ldots,x_k;0) &= \pi_0\tilde{\pi}(x_1,\ldots,x_k), \\ \pi(1,\ldots,1;m) &= \pi_0\rho^m\tilde{\pi}(1,\ldots,1) \quad \text{for } m > 0, \qquad (5.41) \end{aligned}$$

where $\pi_0 > 0$ is a normalizing constant. These formulas in a somewhat different form were obtained in [4].

Queue length: Recall that

$$Q = \lim_{t\to\infty}\frac{1}{t}\int_0^t Q(s)ds \quad \text{and} \quad \bar{Q} = \lim_{t\to\infty}\frac{1}{t}\int_0^t \bar{Q}(s)ds$$

are, respectively, the long-run time averages of the queue length and total queue length of a stable queuing system given in Proposition 3.13, which exist and are non-random. By the discussion in §3.8, for a nonlattice queuing system, they are also equal to the expected queue length and the expected total queue length in steady state, that is,

$$Q = \sum_{j=k}^{\infty}(j-k)\pi_j \quad \text{and} \quad \bar{Q} = \sum_{j=1}^{\infty}j\pi_j,$$

where k is the number of servers and π_j is the steady-state total queue length distribution. For a stable M/M/k, by (5.31) with $\xi = \lambda/\mu$ and $\rho = \lambda/(k\mu)$,

$$\begin{aligned} Q &= \sum_{j=k+1}^{\infty}(j-k)\pi_j = \frac{\xi^k\rho\pi_0}{k!}\sum_{j=1}^{\infty}j\rho^{j-1} \\ &= \frac{\xi^k\rho\pi_0}{k!}\frac{d}{d\rho}\frac{1}{1-\rho} = \frac{\xi^k\rho\pi_0}{k!(1-\rho)^2}. \qquad (5.42) \end{aligned}$$

This agrees with the formula for Q in Theorem 3.17 in the case of $k = 1$. A formula for $\bar{Q}$ in a stable M/M/k will be given in (5.45) later.

For the system with two heterogeneous servers considered earlier, a similar computation based on (5.34) leads to

$$Q = \frac{\lambda \rho^2 \pi_0}{\nu(1-\rho)^2}, \quad \text{where} \quad \rho = \frac{\lambda}{\mu_A + \mu_B}. \tag{5.43}$$

Waiting time: Recall the long-run average waiting time W and the long-run average total waiting time $\bar{W}$ of a stable queuing system were defined in §3.4 as

$$W = \lim_{n \to \infty} \frac{W_1 + \cdots + W_n}{n} \quad \text{and} \quad \bar{W} = \lim_{n \to \infty} \frac{\bar{W}_1 + \cdots + \bar{W}_n}{n},$$

where W_n and $\bar{W}_n$ are, respectively, the successive waiting times (before service) and total waiting times. The W and $\bar{W}$ exist and are non-random. By (3.56) in §3.8, with Poisson arrivals, W and $\bar{W}$ are also equal to, respectively, the expected waiting and total waiting times of a fictitious customer entering the system in steady state.

By Little's formula (3.12), $W = Q/\lambda$ and $\bar{W} = \bar{Q}/\lambda$. Thus, for a stable M/M/k, by (5.42),

$$W = \frac{\xi^k \pi_0}{k! k \mu (1-\rho)^2}. \tag{5.44}$$

As W is equal to the expected waiting time of a fictitious customer entering an M/M/k in steady state, so it may also be computed as

$$\begin{aligned}
W &= \sum_{j=k}^{\infty} \frac{j-k+1}{k\mu} \pi_j = \frac{\xi^k \pi_0}{k! k \mu} \sum_{j=k}^{\infty} (j-k+1)\rho^{j-k} \\
&= \frac{\xi^k \pi_0}{k! k \mu} \frac{d}{d\rho} \frac{1}{1-\rho} = \frac{\xi^k \pi_0}{k! k \mu (1-\rho)^2}.
\end{aligned}$$

Because $\bar{W} = W + 1/\mu$, we obtain $\bar{Q}$ for a stable M/M/k,

$$\bar{Q} = \lambda \bar{W} = \lambda \left(W + \frac{1}{\mu} \right) = Q + \xi = \frac{\xi^k \rho \pi_0}{k!(1-\rho)^2} + \xi. \tag{5.45}$$

This may also be computed directly as the expected total queue length in steady state; see Exercise 5.12.

Exercise 5.11 Consider a 3-server queuing system with exponential arrival rate 5 and exponential service rate 2 for each server.
(a) Assume the system has an infinite holding capacity. Find the long-run fraction of time when there are at least 5 customers in the system.
(b) Now assume the system has a holding capacity of 5. Find the long-run fraction of time when the system is full.
(c) Find the expected waiting time for a fictitious customer who enters the system in steady state, assuming first an infinite holding capacity and then a holding capacity of 5.

Exercise 5.12 For a stable $M/M/k$ with arrival rate λ and service rate μ for each server, use the formula (5.31) for the stationary distribution $\{\pi_j\}$ to directly compute $\bar{Q} = \sum_{j=1}^{\infty} j\pi_j$ as the expected total queue length in steady state, and compare with (5.45).

Exercise 5.13 Customers arrive according to a Poisson process of rate 5 to a 2-server system with different exponential service rates 3 and 4. Assume any customer arriving at an empty system is to be served by the faster server.
(a) Find the long-run fraction of time when the slower server is idle.
(b) Find the expected total queue length $\bar{Q}$ in steady state and hence find the expected total waiting time $\bar{W} = \bar{Q}/\lambda$ by Little's formula.
(c) Use the stationary distribution in Theorem 5.13 to compute the expected total waiting time $\bar{W}$ directly and compare with result in (b).

Exercise 5.14 Consider a stable $M/M/1$ with arrival rate λ and service rate μ. Let Z and $\bar{Z}$ be, respectively, the waiting time and the total waiting time of a fictitious customer entering the system in steady state.
(a) Show that $\bar{Z}$ is exponential of rate $\mu - \lambda$.
(b) Find the distribution function of Z.

Exercise 5.15 ($M/M/1$ with delayed service)   Consider a stable $M/M/1$ with arrival rate λ and service rate μ. Assume a delayed service as in Exercise 3.15 with $m > 1$ customers to start the service.
(a) Identify the states of this system and find its stationary distribution.
(b) Find the expected queue length Q in steady state.
Note: By Exercise 3.15(c) and (5.42), $Q = \rho^2/(1-\rho) + (m-1)/2$, but in (b), compute Q directly using the stationary distribution from (a).

5.6 Time reversibility

Time reversed MC: Let $X(t)$ be a continuous time MC. Let $P_{ij}(t)$ be its transition probability function and let $q_{ij} = P'_{ij}(0)$ be its transition rate from i to j for $i \neq j$. Fix $T > 0$ and define the time-reversed process by

$$X^*(t) = X(T - t) \quad \text{for } 0 \leq t \leq T.$$

We will show that $X^*(t)$ is a continuous time MC. However, a continuous time MC is required to be regular, in particular, it is required to be right continuous in t, whereas $X^*(t)$ just defined is left continuous. Therefore, we will redefine $X^*(t)$ to be $X^*(t+)$. As $X^*(t) = X^*(t+)$ a.s. for any fixed $t > 0$ (because the probability that a transition occurs at time t is zero), we may assume $X^*(t) = X(T - t)$ in any probability computation.

Theorem 5.14 *Let $X(t)$ be irreducible and in steady state under the stationary distribution $\{\pi_j\}$. Then $X^*(t)$ is a continuous time MC with transition probability function*

$$P_{ij}^*(t) = \pi_j P_{ji}(t)/\pi_i. \tag{5.46}$$

Proof: We will prove that $X^*(t)$ has the Markov property with transition probability function given by (5.46). For $0 \leq t_1 < t_2 < \cdots < t_n < t_{n+1} \leq T$,

$$P[X^*(t_{n+1}) = j \mid X^*(t_n) = i, X^*(t_{n-1}) = i_{n-1}, \ldots, X^*(t_1) = i_1]$$
$$= P[\text{future for } X^* \mid \text{present for } X^* \text{ at time } t_n, \text{past for } X^*]$$
$$= P[\text{past for } X \mid \text{present for } X \text{ at time } T - t_n, \text{future for } X]$$
$$= \frac{P[\text{past for } X, \text{present for } X, \text{future for } X]}{P[\text{present for } X, \text{future for } X]}$$
$$= \frac{P[\text{future } X \mid \text{present } X, \text{past } X] P[\text{present } X, \text{past } X]}{P[\text{future } X \mid \text{present } X] P[\text{present } X]}$$
$$= \frac{P[\text{present } X, \text{past } X]}{P[\text{present } X]} \quad \text{(Markov property)}$$
$$= \frac{P[X(T - t_n) = i, X(T - t_{n+1}) = j]}{P[X(T - t_n) = i]}$$
$$= \frac{P[X(T - t_n) = i \mid X(T - t_{n+1}) = j] P[X(T - t_{n+1}) = j]}{P[X(T - t_n) = i]}$$

$$= \frac{P_{ji}(t_{n+1} - t_n)\pi_j}{\pi_i} = \cdots = P[X^*(t_{n+1}) = j \mid X^*(t_n) = i]. \quad \Diamond$$

Transition rates of time-reversed MC: Differentiating (5.46) at $t = 0$ yields the transition rate from i to j of the time-reversed process $X^*(t)$ $(i \neq j)$,

$$q_{ij}^* = \pi_j q_{ji}/\pi_i. \tag{5.47}$$

Note that $X^*(t)$ and $X(t)$ have the same total transition rate from i, that is, $q_i^* = q_i$, because $q_i^* = \sum_{j \neq i} q_{ij}^* = (\sum_{j \neq i} \pi_j q_{ji})/\pi_i = \pi_i q_i/\pi_i = q_i$ by the balance equation at i.

Time reversible MC: The MC $X(t)$ is called time reversible if it has the same distribution as the time-reversed MC $X^*(t)$, that is, if $q_{ij}^* = q_{ij}$. By (5.47), we see that $X(t)$ is time reversible if and only if

$$\pi_i q_{ij} = \pi_j q_{ji}. \tag{5.48}$$

Proposition 5.15 *An irreducible continuous time MC $X(t)$ is time reversible if and only if there is a distribution $\{\pi_j\}$ satisfying (5.48). Then $\{\pi_j\}$ is also its stationary distribution.*

The easy proof of the above proposition is left to Exercise 5.16.

Theorem 5.16 *(Kolmogorov reversibility criterion)* *A continuous time ergodic MC is time reversible if and only if the product of transition rates q_{ij} along any loop of states is equal to that of the reversed loop, that is, for any states $i, i_1, i_2, \ldots, i_n$,*

$$q_{ii_1} q_{i_1 i_2} \cdots q_{i_n i} = q_{ii_n} q_{i_n i_{n-1}} \cdots q_{i_1 i}. \tag{5.49}$$

Proof: Assume the MC is time reversible and has the stationary distribution π_j. It is trivial that (5.49) holds for a loop of two states. For a loop of three states i, j, k, $q_{ij} q_{jk} q_{ki} = \pi_i q_{ij} q_{jk} q_{ki}/\pi_i = q_{ji} \pi_j q_{jk} q_{ki}/\pi_i = q_{ji} q_{kj} \pi_k q_{ki}/\pi_i = q_{ji} q_{kj} q_{ik}$. This proves (5.49) for a loop of three states. Similarly, for a loop of any number of states. Next, assume (5.49) holds for any loop. Write q_{jj} for $-q_j$, a diagonal element of the Q-matrix. Suppose the left-hand side of (5.49) has a factor q_{jk}. Then the right-hand side must have q_{kj}. An extra factor q_{kk} may be inserted on both sides so that the left has $q_{jk} q_{kk}$ and the right has $q_{kk} q_{kj}$. This means that one may allow a loop $i \to i_1 \to i_2 \to \cdots \to i_n \to i$ in (5.49) to have two identical adjacent indices. Then summing over

$i_1, i_2, \ldots, i_{n-1}$ and writing j for i_n yields $(Q^{n-1})_{ij}q_{ji} = q_{ij}(Q^{n-1})_{ji}$, and hence $P_{ij}(t)q_{ji} = (e^{tQ})_{ij}q_{ji} = q_{ij}(e^{tQ})_{ji} = q_{ij}P_{ji}(t)$. Letting $t \to \infty$, we obtain $\pi_j q_{ji} = q_{ij}\pi_i$ and hence the MC is time reversible. $\diamondsuit$

Example 5.17 A BD process in steady state is time reversible. This follows directly from Kolmogorov criterion because moving along any loop is just moving along a line segment back and forth.

Example 5.18 (departure process from M/M/k): Recall that the total queue length $X(t)$ for M/M/k queue at time t is a BD process, and hence is time reversible if the system is stable. Because in the reversed time, arrivals become departures and departures become arrivals, it follows that the departure process $Y(t)$, defined as the number of departures by time t, has the same distribution as the arrival process, that is, a Poisson process of rate λ (arrival rate).

Example 5.19 (Two heterogeneous servers): The case of a stable queuing system with exponential arrival rate λ, two servers with differential exponential service rates μ_A and μ_B, and probabilities p and $q = 1 - p$ of assigning server A and B to an arrival at an empty queue, is considered in §5.5. By the transition graph displayed there with states $0, 1A, 1B, 2, 3, \ldots$, it is easy to see that to use Kolmogorov's criterion to check time reversibility, it suffices to look at the loop $0 \to 1A \to 2 \to 1B \to 0$. The corresponding equation is

$$(\lambda p)\lambda\mu_A\mu_B = (\lambda q)\lambda\mu_B\mu_A.$$

It follows that the system is time reversible if and only if $p = q = 1/2$.

Two queuing systems in tandem Customers arrive according to a Poisson process of rate λ at queuing system 1 with k_1 servers of the same exponential service rate μ_1, and then enter queuing system 2 with k_2 servers of the same exponential service rate μ_2. The states of the combined system are pairs (i, j), where i and j are, respectively, the total queue lengths at systems 1 and 2. Let $X(t) = (X_1(t), X_2(t))$ be the state of the combined system at time t. Because the time between the change of states is exponential, $X(t)$ is a continuous time MC. Its transition rates may be written down and from them one may solve for the stationary distribution.

If system 1 is in steady state, then by the time reversibility, the departures from system 1 form a Poisson process of rate λ, and hence system 2 becomes $M/M/k_2$. Although the two processes $X_1(t)$ and $X_2(t)$ are not independent, but in steady state, they are independent at any fixed time t, as stated in the next theorem. This allows us to obtain the stationary distribution of $X(t)$ more quickly in terms of the stationary distributions of $X_1(t)$ and $X_2(t)$.

Theorem 5.20 *Assume $\lambda/(k_1\mu_1) < 1$ and $\lambda/(k_2\mu_2) < 1$.*
(a) Then $X(t)$ is an ergodic MC, and in steady state, $X_1(t)$ and $X_2(t)$ are independent at any time t. Moreover, the stationary distribution $\{\pi_{ij}\}$ of $X(t)$ is given by
$$\pi_{ij} = \pi_i^1 \pi_j^2, \tag{5.50}$$
where $\{\pi_j^\alpha\}$, for $\alpha = 1$ or 2, is the stationary distribution of the total queue length of an $M/M/k_\alpha$ with arrival rate λ and service μ_α for each of k_α servers.
(b) Let $\bar{W}^1$ and $\bar{W}^2$ be, respectively, the total waiting times of a customer in system 1 and system 2. Then in steady state, $\bar{W}^1$ and $\bar{W}^2$ are independent.

Proof: As mentioned earlier, when system 1 is in steady state, system 2 is $M/M/k_2$ with arrival rate λ. The $X_2(t)$ is determined by the departures from system 1 before time t, and in the reversed time, those departures become arrivals after time t and so are independent of $X_1(t)$. We now have, in steady state,

$$\begin{aligned} P[X(t) = (i,j)] &= P[X_1(t) = i, X_2(t) = j] \\ &= P[X_1(t) = i]P[X_2(t) = j] = \pi_i^1 \pi_j^2. \end{aligned}$$

Then the above, denoted as π_{ij}, is the stationary distribution of $X(t)$. Because $X(t)$ is clearly irreducible, it follows that $X(t)$ is ergodic. This proves (a).

The total waiting time of a customer in a queuing system may be determined by observing the change in the total queue length as it starts with his arrival (an increase in queue length) and it ends after n departures (decreases in queue length), where n is the total queue length at the time of the arrival. This description holds also in the reversed time. The $\bar{W}^2$ is determined by the departures from system 1 before this customer leaves system 1, but in the reversed time these departures become arrivals after he has entered 1 from 2, which is independent of the time he will spend in system 1. This proves (b). $\diamondsuit$

Departure from heterogeneous servers: For an exponential queuing system with two heterogeneous servers as discussed in §5.5, the total queue length is a BD process when states $1A$ and $1B$ are combined, and hence is time reversible in steady state. The departure process thus is equal in distribution to the arrival process. It follows that the total queue length of this queue and that of a second queue in tandem are independent at any time t.

Example 5.21 (two servers in tandem): Customers arrive at a system according to a Poisson process of rate λ to be served by two servers in tandem with independent exponential service rates μ_1 and μ_2, respectively. The system is described by an irreducible continuous time MC with states (i,j), $i,j = 0,1,2,3,\ldots$. By Theorem 5.20, if $\lambda < \mu_1$ and $\lambda < \mu_2$, then the stationary distribution exists and is given by

$$\pi_{i,j} = (\frac{\lambda}{\mu_1})^i(1-\frac{\lambda}{\mu_1})(\frac{\lambda}{\mu_2})^j(1-\frac{\lambda}{\mu_2}) \quad \text{for } i,j = 0,1,2,3,\ldots. \quad (5.51)$$

This may also be obtained by solving the balance equations:

$$\begin{aligned}
\lambda\pi_{0,0} &= \mu_2\pi_{0,1}, \\
(\lambda+\mu_1)\pi_{i,0} &= \lambda\pi_{i-1,0} + \mu_2\pi_{i,1} \quad \text{(for } i > 0), \\
(\lambda+\mu_2)\pi_{0,j} &= \mu_1\pi_{1,j-1} + \mu_2\pi_{i,j+1} \quad \text{(for } j > 0), \\
(\lambda+\mu_1+\mu_2)\pi_{i,j} &= \lambda\pi_{i-1,j} + \mu_1\pi_{i+1,j-1} + \mu_2\pi_{i,j+1} \quad \text{(for } i,j > 0).
\end{aligned}$$

It will require some effort to solve these equations, but it is easier to verify that $\pi_{i,j}$ given in (5.51) satisfy these equations.

Example 5.22 (two servers in tandem with finite holding capacities): As in Example 5.21, but now assume both servers 1 and 2 have finite holding capacities N_1 and N_2, and when server 2 is full, a customer ending his service at server 1 will stay with server 1 and thus block its service to next customer. Server 1 resumes when a space at server 2 becomes available.

The states of the system may be represented by (i,j) with i being the number at server 1 and j the number at server 2, but a special mark should be used to indicate whether server 1 is blocked, say, letter b is inserted if server 1 is blocked. For example, with $N_1 = 3$ and $N_2 = 2$, all the states with transition rates are depicted in the following transition diagram.

$$
\begin{array}{cccccccc}
(0,0) & \lambda \to & (1,0) & \lambda \to & (2,0) & \lambda \to & (3,0) \\
\uparrow \mu_2 & \swarrow \mu_1 & \uparrow \mu_2 & \swarrow \mu_1 & \uparrow \mu_2 & \swarrow \mu_1 & \uparrow \mu_2 \\
(0,1) & \lambda \to & (1,1) & \lambda \to & (2,1) & \lambda \to & (3,1) \\
\uparrow \mu_2 & \swarrow \mu_1 & \uparrow \mu_2 & \swarrow \mu_1 & \uparrow \mu_2 & \swarrow \mu_1 & \uparrow \mu_2 \\
(0,2) & \lambda \to & (1,2) & \lambda \to & (2,2) & \lambda \to & (3,2) \\
& \nwarrow \mu_2 & \downarrow \mu_1 & \nwarrow \mu_2 & \downarrow \mu_1 & \nwarrow \mu_2 & \downarrow \mu_1 \\
& & (1b,2) & \lambda \to & (2b,2) & \lambda \to & (3b,2)
\end{array}
$$

Exercise 5.16 Prove Proposition 5.15.

Exercise 5.17 Consider two queues in tandem. The first queue has two servers with different exponential service rates 2 and 3, and the second queue has a single server with exponential service rate 5. The customers arrive to the first queue at an exponential rate 4 and when the queue is empty, the arriving customer is assigned to either server with same probability. Assume both queues have infinite holding capacity. Find the long-run fraction of time when both queues are empty.

Exercise 5.18 Customers arrive according to a Poisson process of rate 5 at a service station with two servers in tandem.
(a) Assume both servers have the same exponential service rate 6 and infinite holding capacity. Find the long-run fraction of time when server 1 is free but at least 3 customers are with server 2.
(b) Now assume servers 1 and 2 have different exponential service rates 2 and 3, respectively, and the same holding capacity of 1. Find the long-run fraction of time when server 1 is blocked.

Exercise 5.19 The notion of time reversibility can be applied to a discrete time MC. Let X_n be an irreducible discrete time MC under stationary distribution π_j, and let p_{ij} be its transition probabilities. Fix $N > 0$ and define the time-reversed process X_n^* by $X_n^* = X_{N-n}$ for $0 \le n \le N$.
(a) Show that X_n^* is a discrete time MC with transition probabilities $p_{ij}^* = \pi_j p_{ji} / \pi_i$.
(b) The MC X_n is called time reversible if it has the same distribution as the time-reversed process X_n^*, that is, if $p_{ij}^* = p_{ij}$. Prove the Kolmogorov criterion: X_n is time reversible if and only if the product of transition probabilities p_{ij} along any loop of states is equal to that along the reversed loop.

Exercise 5.20 Consider a system of two exponential servers in tandem with service rates $\mu_1 = 3$ and $\mu_2 = 4$, respectively. The arrival process to the first queue is Poisson of rate $\lambda = 2$. Find the probability that a customer entering the system in steady state will spend no more than 1 unit of time in system.

5.7 Hitting time and phase-type distributions

Hitting time of a set: Let $X(t)$ be an irreducible continuous time MC and let A be a collection of its states. We are interested in finding the distribution of the first time τ when the MC enters A from a state outside, where τ, called the hitting time of A, is formally defined by

$$\tau = \inf\{t > 0; \ X(t) \in A\}. \tag{5.52}$$

It is set to ∞ if $X(t) \notin A$ for all $t > 0$. The probability of hitting a particular state in A when A is hit may be obtained by working with the embedded discrete time MC as in §4.6.

Since we are only concerned with the behavior of the MC up to the time when it first enters A, we may and will assume that the states in A are absorbing, that is, $q_i = 0$ for all $i \in A$. We will also assume both A and its complement A^c consist of finitely many states, and from each state in A^c, A is accessible. Then all the states in A are transient and hence τ is a.s. finite.

Q-matrix in block form: With A and A^c denoting, respectively, the sets of absorbing and non-absorbing states, the Q-matrix in (5.11) may be written in the following block form, which is useful in performing various matrix computations in the sequel,

$$Q = \begin{bmatrix} T & T_2 \\ 0 & 0 \end{bmatrix} \begin{matrix} A^c \\ A \end{matrix}, \tag{5.53}$$
$$ \begin{matrix} A^c & A \end{matrix}$$

where the sub-matrices T and T_2 contain transition rates from non-absorbing states to non-absorbing states and to absorbing states respectively. Note that T is a square matrix.

Transition probability matrix: By (5.13),

$$\mathbf{P}(t) = e^{tQ} = \begin{bmatrix} e^{tT} & T^{-1}(e^{tT} - I)T_2 \\ 0 & I \end{bmatrix}, \tag{5.54}$$

where I denotes an identity matrix of a suitable size. Note that inverse matrix T^{-1} above exists, because if not, then there must be a non-zero row vector b such that $bT = 0$, and hence $be^{tT} = b$. This is impossible because $P_{ij}(t) \to 0$ as $t \to \infty$ for transient states i and j.

Distribution function of the hitting time τ: Let p be the row vector of the initial distribution of the Markov chain $X(t)$ on A^c. Then

$$
\begin{aligned}
F_\tau(t) &= P(\tau \le t) = 1 - P(\tau > t) = 1 - \sum_{i,j \in A^c} p_i P_{ij}(t) \\
&= 1 - p\, e^{tT}[1 \ldots 1]', \tag{5.55}
\end{aligned}
$$

where $[1 \ldots 1]'$ is the column vector of 1's of a suitable size (here the prime $'$ denotes the matrix transpose).

Probability density function of τ:

$$f_\tau(t) = \frac{d}{dt}F_\tau(t) = -pTe^{tT}[1 \ldots 1]'. \tag{5.56}$$

The computations done here and in the sequel are routine operations on real-valued functions, but they are also valid on matrix functions.

Laplace transform of τ:

$$
\begin{aligned}
L_\tau(s) &= E[e^{-s\tau}] = \int_0^\infty e^{-st} f_\tau(t)dt = -pT\left(\int_0^\infty e^{-st}e^{tT}dt\right)[1 \ldots, 1]' \\
&= pT(-sI + T)^{-1}[1 \ldots 1]'. \tag{5.57}
\end{aligned}
$$

The above holds when $|s|$ is sufficiently small, and note that the inverse matrix $(-sI + T)^{-1}$ exists for small $|s|$ because T^{-1} exists.

The matrix integral in (5.57) is evaluated formally noting that $(-sI + T)^{-1}e^{(-sI+T)t}$ is an antiderivative of the matrix-valued function $e^{(-sI+T)t}$ in t. To justify this computation, we need to show $\lim_{t\to\infty} e^{t(-sI+T)} = 0$. Because A^c is transient, $\lim_{t\to\infty} e^{tT} = 0$. This implies that all the eigenvalues of T have negative real parts. Then the same must be true when T is replaced by $-sI + T$ for small $|s|$ and hence $\lim_{t\to\infty} e^{t(-sI+T)} = 0$.

Mean and variance:

$$E(\tau) = -L'_\tau(0) = -pT^{-1}[1 \ldots 1]', \tag{5.58}$$
$$\mathrm{Var}(\tau) = L''_\tau(0) - L'_\tau(0)^2 = 2pT^{-2}[1 \ldots 1]' - (E\tau)^2, \tag{5.59}$$

where $T^{-2} = (T^{-1})^2$.

Example 5.23 Consider an M/M/2 system with arrival rate 5 and service rate 3 for each of two servers. Let τ be the first time that there are 3 customers in system. Find $E(\tau)$ and $P(\tau > 2)$ assuming the system started with 0 customer.

Solution: This may be modeled by a Markov chain on states 0, 1, 2, and 3 with state 3 being absorbing, obtained from the total queue length process of M/M/2 by setting state 3 absorbing. The T-matrix (transition among states $0, 1, 2$) and its inverse are

$$T = \begin{bmatrix} -5 & 5 & 0 \\ 3 & -8 & 5 \\ 0 & 6 & -11 \end{bmatrix}, \quad T^{-1} = \begin{bmatrix} -0.4640 & -0.4400 & -0.2000 \\ -0.2640 & -0.4400 & -0.2000 \\ -0.1440 & -0.2400 & -0.2000 \end{bmatrix}.$$

The initial distribution vector is $p = [1\ 0\ 0]$. Then using MATLAB®,

$$E(\tau) = -(1\ 0\ 0)T^{-1}(1\ 1\ 1)' = 1.1040,$$
$$P(\tau > 2) = [1\ 0\ 0]e^{2T}[1\ 1\ 1]' = 0.1382.$$

Phase-type distributions: A nonnegative-valued random variable is said to have a phase-type distribution if it has the same distribution as the hitting time τ of a set A of absorbing states of an irreducible continuous time MC as described above. The MC is assumed to have finitely many transient states, called phases. Such a distribution is determined by a matrix T, which is a part of the Q-matrix as in (5.53), and a row vector p of initial distribution on phases. Its distribution function F_τ, pdf f_τ, Laplace transform L_τ, mean $E(\tau)$, and variance $\mathrm{Var}(\tau)$ are given by (5.55) through (5.59).

An Erlang distribution $\mathrm{Erlang}(n, \lambda)$ is a phase-type distribution that has $n - 1$ phases in a series, with

$$T = \begin{bmatrix} -\lambda & \lambda & 0 & \cdots & 0 & 0 \\ 0 & -\lambda & \lambda & \cdots & 0 & 0 \\ \cdots & \cdots & \cdots & \cdots & \cdots & \cdots \\ 0 & 0 & 0 & \cdots & 0 & -\lambda \end{bmatrix} \quad \text{and} \quad p = \begin{pmatrix} 1 \\ 0 \\ \cdots \\ 0 \end{pmatrix}.$$

Because the transitions between phases are exponential, a queuing system with phase-type distributions for inter-arrival and service times may be modeled by a continuous time MC. From example, consider a single-server system with no waiting space. The inter-arrival and service time distributions are phase type with the following T-matrices and p-vectors:

$$T^a = \begin{bmatrix} -a - h & a \\ b & -b - k \end{bmatrix}, \quad p^a = \begin{pmatrix} u \\ v \end{pmatrix},$$

and

$$T^s = \begin{bmatrix} -\alpha - \eta & \alpha \\ \beta & -\beta - \kappa \end{bmatrix}, \quad p^s = \begin{pmatrix} \mu \\ \nu \end{pmatrix}.$$

There are six states, which may be labeled as 01, 02, 111, 112, 121, and 122, where for each state, the first digit is the number of customers in system, the second is the phase of arrive time, and the third is the phase of service time, noting there is no service when there is no customer in system. The transition rates are depicted below with states 01 and 02 repeated for easy depiction.

$$
\begin{array}{cccccccccccc}
(01) & \overset{h u \mu}{\rightarrow} & \overset{\eta}{\leftarrow} & (111) & \overset{\alpha}{\rightarrow} & \overset{\beta}{\leftarrow} & (112) & \overset{\kappa}{\rightarrow} & \overset{h u \nu}{\leftarrow} & (01) \\
 & \overset{h v \mu}{} & & & & & & & \overset{h v \nu}{} & \\
a \downarrow & \searrow & & a \downarrow & & & a \downarrow & & \swarrow & \\
 & \overset{k u \mu}{} & & & & & & & \overset{k u \nu}{} & \\
b \uparrow & \nearrow & & b \uparrow & & & b \uparrow & & \nwarrow & \\
(02) & \overset{k v \mu}{\rightarrow} & \overset{\eta}{\leftarrow} & (121) & \overset{\alpha}{\rightarrow} & \overset{\beta}{\leftarrow} & (122) & \overset{\kappa}{\rightarrow} & \overset{k v \nu}{\leftarrow} & (02).
\end{array}
$$

The study of a queuing system with phase-type distribution is helpful in understanding a general queuing system, because any distribution on $\mathbb{R}_+ = [0, \infty)$ may be approximated arbitrarily close by a phase type distribution in some sense (see [1, Section III.4]).

Exercise 5.21 In Exercise 5.6.2 (b) (two servers in tandem with Poisson arrival of rate 5, exponential service rates 2 and 3, and holding capacity 1 at both servers), find the mean and standard deviation for the first time when server 1 is blocked assuming the system started empty. Also find the probability that the first blocking time is between 0.5 and 5.

Exercise 5.22 In a single-server queuing system, the inter-arrival time has a phase-type distribution with rate matrix T and initial vector p given by

$$T = \begin{bmatrix} -3 & 1 \\ 2 & -3 \end{bmatrix} \quad \text{and} \quad p = \begin{bmatrix} 0.6 \\ 0.4 \end{bmatrix},$$

and service time is Erlang$(2, 1/3)$. Assume that when the server is busy, no customer enters the system. Find the long-run fraction of the time when the server is busy.

5.8 Queuing systems with time-varying rates

Time-varying arrival and service rates: Consider a queuing system to which customers arrive according to a nonhomogeneous Poisson process of rate $\lambda(t)$ as defined in § 2.6. The system has k homogeneous servers, and the number of services each server can perform forms a nonhomogeneous Poisson process of rate $\mu(t)$. Thus, the number X of arrivals in a time interval (a, b) is a Poisson random variable of mean $\int_a^b \lambda(t)dt$, and the number Y_i of services server i can perform is also Poisson of mean $\int_a^b \mu(t)dt$. Moreover, $X, Y_1, Y_2, \ldots, Y_k$ are independent, and are also independent of what happens before time a. Note that the actual number of services performed by the ith server may be less than Y_i due to possible idle time.

This is not a queuing system as discussed in Chapter 3 because the arrivals do not follow a renewal process, but may provide a more realistic model for queues because it allows the arrival rate and service rate to change in time. Such a system is denoted by $M(t)/M(t)/k$. When there is a finite holding capacity $N \geq k$, it will be denoted as $M(t)/M(t)/k/N$.

Distribution of total queue length: Let $X(t)$ be the total queue length at time t, that is, the number of customers in the system at time t. It can be shown that $X(t)$ is a continuous time MC, but not time homogeneous. Let

$$P_n(t) = P[X(t) = n], \quad n = 0, 1, 2, \ldots.$$

Consider an $M(t)/M(t)/k/N$ system. Note that

$$P[\text{two or more events during } (t, \ t+h)] \ = \ o(h),$$

where an event means an arrival or a service completion, and $o(h)$ denotes a function satisfying $o(h)/h \to 0$ as $h \to 0$. By the total probability law, $P_n(t + h)$ is equal to

$$
\begin{aligned}
&P_{n-1}(t)P[X(t+h) - X(t) = 1 \mid X(t) = n - 1] \\
+\; &P_n(t)P[X(t+h) - X(t) = 0 \mid X(t) = n] \\
+\; &P_{n+1}(t)P[X(t+h) - X(t) = -1 \mid X(t) = n + 1] + o(h),
\end{aligned}
$$

setting $P_{-1}(t) = P_{N+1}(t) = 0$. By the independent increments of non-homogeneous Poisson arrivals and services, we may compute

$$
P_n'(t) = \lim_{h \to 0} \frac{1}{t}[P_n(t + h) - P_n(t)]
$$

to obtain the following system of differential equations:

$$
\begin{aligned}
P_0'(t) &= -\lambda(t)P_0(t) + \mu(t)P_1(t) \\
P_n'(t) &= \lambda(t)P_{n-1}(t) - [\lambda(t) + n\mu(t)]P_n(t) + (n+1)\mu(t)P_{n+1}(t) \\
&\quad \text{(for } 1 \leq n < k) \\
P_n'(t) &= \lambda(t)P_{n-1}(t) - [\lambda(t) + k\mu(t)]P_n(t) + k\mu(t)P_{n+1}(t) \\
&\quad \text{(for } k \leq n < N) \\
P_N'(t) &= \lambda(t)P_{N-1}(t) - k\mu(t)P_N(t). \tag{5.60}
\end{aligned}
$$

Let $P_{\cdot}(t)$ and $P'(t)$ be, respectively, the column vectors with components $P_n(t)$ and $P_n'(t)$ for $0 \leq n \leq N$, and let $Q(t)$ be the $(N+1) \times (N+1)$ matrix formed by the coefficients, aligned by the index n in $P_n(t)$, on the right-hand side of (5.60). Then the equations in (5.60) may be written more concisely in the following matrix form:

$$
P'(t) = Q(t)P_{\cdot}(t). \tag{5.61}
$$

Assume $\lambda(t)$ and $\mu(t)$ are continuous. By the general theory of differential equations, the system (5.60) has a unique solution under any initial condition. For an initially empty queue, the initial condition is: $P_0(0) = 1$ and $P_n(0) = 0$ for $1 \leq n \leq N$.

Example 5.24 For an M(t)/M(t)/2/4 system with arrival rate $\lambda(t) = 2t(5 - t)$, $0 \leq t \leq 5$, and service rate $\mu(t) = 0.2t$ for each server, use MATLAB® to find and plot the expected total queue length $E[X(t)]$ for $0 \leq t \leq 5$, assuming the system is empty at time $t = 0$.

Solution: In this example, the system (5.60) is

$$
\begin{aligned}
P_0'(t) &= -2t(5-t)P_0(t) + 0.2tP_1(t) \\
P_1'(t) &= 2t(5-t)P_0(t) - [2t(5-t) + 0.2t]P_1(t) + 2(0.2t)P_2(t) \\
P_2'(t) &= 2t(5-t)P_1(t) - [2t(5-t) + 2(0.2t)]P_2(t) + 2(0.2t)P_3(t) \\
P_3'(t) &= 2t(5-t)P_2(t) - [2t(5-t) + 2(0.2t)]P_3(t) + 2(0.2t)P_4(t) \\
P_4'(t) &= 2t(5-t)P_3(t) - 2(0.2t)P_4(t).
\end{aligned}
$$

They may be solved numerically using the MATLAB® function ode45 under the initial condition $P_0(0) = 1$ and $P_n(0) = 0$ for $1 \le n \le 4$. The expected total queue length is then computed as $E[X(t)] = \sum_{n=0}^{4} nP_n(t)$. The result is plotted in Figure 5.1.

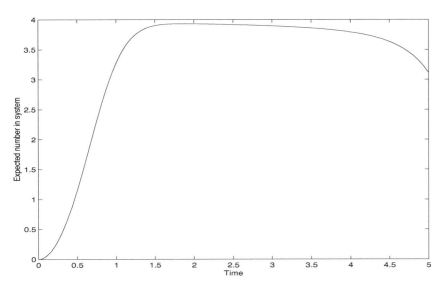

FIGURE 5.1: The expected total queue length in M(t)/M(t)/2/4

A fast queuing system: We now consider an M(t)/M(t)/k system with fast time-varying arrival and service rates. In such a system, because of high rates, the system may reach steady state at any time when the arrival rate is less than the total service rate in an average sense in the preceding period. This allows us to approximate the system at such a time using steady-state distribution.

We will assume the arrival rate $\lambda(t)$ and the service rate $\mu(t)$ of each server are piecewise continuous, and we will allow the number $k = k(t)$ of servers to change so that $k(t)$ is a step function. Thus, any finite time interval may be divided into several subintervals so that $\lambda(t)$ and $\mu(t)$ are continuous, and $k(t)$ is constant, inside each subinterval.

We are interested in the distribution of the total queue length $X(t)$ when $\lambda(t)$ and $\mu(t)$ are large. Precisely, for a small $\varepsilon > 0$, let $X^{\varepsilon}(t)$ be the total queue length of such a system with arrival rate $\lambda(t)/\varepsilon$ and serve rate $\mu(t)/\varepsilon$, called an ε-accelerated system, we want to find $\lim_{\varepsilon \to 0} P[X^{\varepsilon}(t) = n]$ for $n \geq 0$ as an approximation for the total queue length distribution with high arrival and service rates.

Let

$$\rho^*(t) = \sup_{0 \leq r < t} \frac{\int_r^t \lambda(s)ds}{\int_r^t k(s)\mu(s)ds}. \tag{5.62}$$

Note that $\rho^*(t) < 1$ means that during any time interval ending at t, the expected number of arrivals is less than the expected number of services the system is capable to perform, whereas $\rho^*(t) > 1$ means that for some interval ending at t, the expected number of arrivals exceeds the expected number of services the system is capable to handle. The function $\rho^*(t)$ may be written as $\rho^*(t; \lambda, \mu, k)$ to indicate its dependence on the functions $\lambda(t), \mu(t)$ and $k(t)$.

Let π_j, $j \geq 0$, be the stationary distribution for the total queue length of a stable M/M/k system with arrival rate λ and service rate μ ($\lambda < k\mu$) given in (5.31). We may write $\pi_j(\lambda, \mu, k)$ for π_j to indicate its dependence on λ, μ, k.

Theorem 5.25 *Fix $t > 0$. Assume λ, μ, k are continuous at t. If $\rho^*(t; \lambda, \mu, k) < 1$, then $\lambda(t) < k(t)\mu(t)$ and*

$$\lim_{\varepsilon \to 0} P[X^{\varepsilon}(t) = n] = \pi_n(\lambda(t), \mu(t), k(t)) \quad \text{for } n = 0, 1, 2, \ldots. \tag{5.63}$$

If $\rho^*(t; \lambda, \mu, k) > 1$, then $X^{\varepsilon}(t) \to \infty$ in distribution as $\varepsilon \to 0$.

The case of $k = 1$ is proved in [8] by a purely analytical method. The general case is established in [3] by a more intuitive probabilistic method, which is outlined below.

For a small $h > 0$, λ, μ, k may be regarded as constants $\lambda(t), \mu(t), k(t)$ in the interval $[t - h, t]$. Running the ε-accelerated system in this small interval has the same effect as running M/M/$k(t)$ with constant rates $\lambda(t)$ and $\mu(t)$ in an interval of length h/ε. As $\varepsilon \to 0$, this interval becomes infinitely long and so the system should reach steady state at time t to have the stationary distribution $\pi_n(\lambda(t), \mu(t), k(t))$. However, for this argument to work, there cannot be a very long queue at the beginning of the interval, and therefore, the condition $\rho^*(t) < 1$ should be imposed. On the other hand, if $\rho^*(t) > 1$, then there is $r < t$ such that $\int_r^t \lambda(s)ds > \int_r^t k(s)\mu(s)ds$, and hence the expected number of customers still in the ε-accelerated system at time t is at least equal to $\int_r^t [\lambda(s) - k(s)\mu(s)]ds/\varepsilon$, which $\to \infty$ as $\varepsilon \to 0$. See [3] for more details.

Example 5.26 Consider a single-server system with Poisson arrivals at rate a in time intervals $[0, 1]$ and $[2, \infty)$, and at rate $4a$ in $(1, 2)$. The server works at a constant exponential rate $2a$. Then the traffic intensity is a function of time and is given by

$$\rho(t) = \begin{cases} 1/2, & 0 \le t \le 1, \\ 2, & 1 < t < 2, \\ 1/2, & t \ge 2. \end{cases}$$

It is easy to see that $\rho^*(t) = \rho(t) = 1/2$ for $t \le 1$ and $\rho^*(t) = \rho(t) = 2$ for $1 < t \le 2$. For $t > 2$, after a little thought, which can be verified by a simple computation, shows that the function $f(r), 0 \le r \le t$, defined by

$$f(r) = \frac{\int_r^t \lambda(s)ds}{\int_r^t \mu(s)ds},$$

is maximized at $r = 1$ with the maximal value $[4a+a(t-2)]/[2a(t-1)] = (2+t)/(2t-2)$. It follows that

$$\rho^*(t) = \begin{cases} 1/2, & 0 \le t \le 1, \\ 2, & 1 < t \le 2, \\ (t+2)/(2t-2), & t > 2. \end{cases}$$

See Figure 5.2 for the graphs of $\rho(t)$ and $\rho^*(t)$. From this, one sees that the rush hour starts at time 1 and ends at time 2, during which the traffic intensity is 2, but $\rho^*(t)$ will remain > 1 until time 4, and hence the congestion will last until time 4, even if the traffic intensity is reduced to $1/2$ after time 2.

Figure 5.3 shows the plot produced by a simple MATLAB® program that computes the expected total queue length in the case of $a = 50$, with holding capacity $N = 100$, which clearly shows the above mentioned congestion pattern.

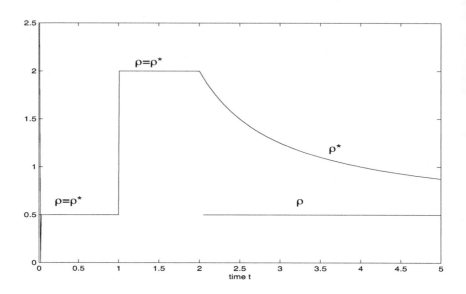

FIGURE 5.2: Graphs of $\rho(t)$ and $\rho^*(t)$.

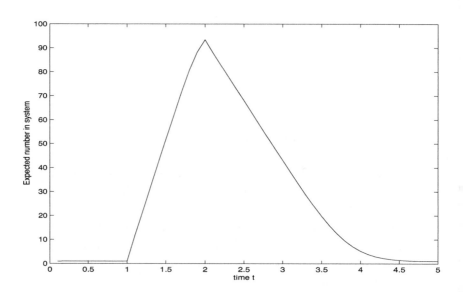

FIGURE 5.3: Expected total queue length.

Chapter 6

Brownian motion and beyond

Brownian motion is probably the most important stochastic process in probability theory and applications. Some of its basic properties are developed in the first half of this chapter, together with a brief discussion of martingales. Based on Brownian motion, the stochastic integrals and stochastic differential equations are introduced, including an application in option pricing in a single stock market. This topic is more advanced than the rest of these notes. An effort is made to present a complete description without being bogged down in advanced theory. For a more complete treatment, the reader is referred to a standard text on stochastic differential equations such as [7].

6.1 Brownian motion

Some motivation: Brownian motion has been used to describe the erratic motion of a pollen particle suspended in a liquid that is caused by the bombardment of surrounding molecules. Viewing the motion in one dimension, one may imagine a particle moving along the real line $\mathbb{R}$ in a continuous random path. It is reasonable to assume that the displacements of the particle over non-overlapping time intervals are independent and the distribution of each displacement depends only on how long the interval is. For any time t, let the interval $[0, t]$ be divided into many small intervals of equal length, then the displacement over $[0, t]$ is the sum of the iid displacements over the small intervals. Suggested by the CLT, the position of the particle at time t should be a normal random variable. This leads to the following formal definition.

Definition of Brownian motion: Let $Z(t)$ be a real-valued process indexed by continuous time $t \geq 0$. It will be called a BM (Brownian motion) if

(i) its paths are continuous;

(ii) it has independent and stationary increments in the sense that for any $s < t$, $Z(t) - Z(s)$ is independent of the process Z up to time s and its distribution depends only on $t - s$;

(iii) $Z(t)$ has a normal distribution for each $t > 0$.

In fact, by the theorem below, which is not proved here but see [5, Theorem 13.4], (iii) is essentially a consequence of (i) and (ii).

Theorem 6.1 *Let $Z(t)$ be a real-valued process satisfying (i) and (ii). Assume for some $s > 0$, $Z(s)$ is not a constant. Then $Z(t)$ is a BM.*

Drift and variance (coefficients): Let $Z(t)$ be a BM. Then $Z(1)$ is normal with mean μ and variance σ^2. By (ii), it is easy to see that for any $s < t$, $Z(t) - Z(s)$ is normal of mean $\mu(t-s)$ and variance $\sigma^2(t-s)$. The constants μ and σ^2 in Theorem 1 will be called, respectively, the drift and variance (coefficients) of the BM.

The distribution of the BM $Z(t)$ as a process is completely determined by the initial position $Z(0)$, the drift μ, and the variance σ^2. In particular, if $Z(0) = z$ (a constant), then for each fixed $t > 0$, the distribution of $Z(t)$ is normal of mean $z + \mu t$ and variance $\sigma^2 t$.

Scaling properties of BM: It is easy to show that if $Z(t)$ is a BM with drift μ and variance σ^2, then for any constants a and $b > 0$, $aZ(bt)$ is a BM with drift $ab\mu$ and variance $a^2 b\sigma^2$. In particular, $-Z(t)$ is a BM with drift $-\mu$ and same variance σ^2.

Time and space shift: Let $Z(t)$ be a BM. It is easy to show that for any fixed $T > 0$, the time-shifted process $Z^T(t) = Z(T+t) - Z(T)$ is a BM with the same drift and variance, and zero initial position. Moreover, $Z^T(t)$ is independent of process $Z(t)$ up to time T.

It can be shown that this property holds also when the constant time shift T is replaced by a stopping time τ, that is, if τ is a stopping time of the BM $Z(t)$, then under the conditional probability $P(\cdot \mid \tau < \infty)$, $Z^\tau(t) = Z(\tau + t) - Z(\tau)$ is a BM with the same drift and variance as $Z(t)$, and zero initial position, which is independent of $Z(t)$ up to τ.

It is easy to see that the BM $Z(t)$ also possesses a space shift property: For any real z, $Z(t) + z$ is a BM with the same drift and variance, but a different starting position $Z(0) + z$.

Standard Brownian motion: An SBM (standard Brownian motion) is a BM $B(t)$ with $B(0) = 0$, drift $\mu = 0$, and variance $\sigma^2 = 1$. Note that if $Z(t)$ is a BM with arbitrary drift μ and variance σ^2, then

$$Z(t) = Z(0) + \sigma B(t) + \mu t, \qquad (6.1)$$

where $B(t) = [Z(t) - Z(0) - \mu t]/\sigma$ is an SBM independent of $Z(0)$.

Markov property: A Brownian motion $Z(t)$ has the Markov property in the following form: For any $t > 0$ and real number z,

$$P[H_t \mid G_t, Z(t) = z] \;=\; P[H_t \mid Z(t) = z] \;=\; P[H_0 \mid Z(0) = z], \quad (6.2)$$

where G_t is an event determined by the BM before time t, such as

$$G_t = [Z(t_1) \in I_1, \ldots, Z(t_k) \in I_k]$$

for $0 \le t_1 < \cdots < t_k \le t$ and intervals $I_1, \ldots, I_k$, H_t is an event determined by the BM after time t, such as

$$H_t = [Z(t + h_1) \in J_1, \ldots, Z(t + h_m) \in J_m]$$

for $0 \le h_1 < \cdots < h_m$ and intervals $J_1, \ldots, J_m$, and H_0 is the event H_t time shifted backward by t, that is, $H_0 = [Z(h_1) \in J_1, \ldots, Z(h_m) \in J_m]$ when H_t is given above.

To prove (6.2), by the independent and stationary increments of $Z(t)$,

$$
\begin{aligned}
& P[Z(t + h_1) \in J_1, \ldots, Z(t + h_m) \in J_m \mid Z(t_1) \in I_1, \\
& \qquad \ldots, Z(t_k) \in I_k, Z(t) = z] \\
=\; & P[Z(t + h_1) - Z(t) \in J_1 - z, \ldots, Z(t + h_m) - Z(t) \in J_m - z \mid \\
& \qquad Z(t_1) \in I_1, \ldots, Z(t_k) \in I_k, Z(t) = z] \\
=\; & P[Z(t + h_1) - Z(t) \in J_1 - z, \ldots, Z(t + h_m) - Z(t) \in J_m - z] \\
& \text{(independent increments)} \\
=\; & P[Z(h_1) - Z(0) \in J_1 - z, \ldots, Z(h_m) - Z(0) \in J_m - z] \\
& \text{(stationary increments)} \\
=\; & P[Z(h_1) \in J_1, \ldots, Z(h_m) \in J_m \mid Z(0) = z] \\
=\; & P[Z(t + h_1) \in J_1, \ldots, Z(t + h_m) \in J_m \mid Z(t) = z].
\end{aligned}
$$

Strong Markov property: A BM $Z(t)$ in fact has the strong Markov property (not proved here) in the sense that the constant t in (6.2) may be replaced by a stopping time τ:

$$
\begin{aligned}
P[H_\tau \mid G_\tau, Z(\tau) = z, \tau < \infty] &= P[H_\tau \mid Z(\tau) = z, \tau < \infty] \\
&= P[H_0 \mid Z(0) = z]. \quad (6.3)
\end{aligned}
$$

Transition density: Let $Z(t)$ be a BM with drift μ and variance σ^2. Because given $Z(0) = z$, $Z(t)$ has a normal distribution of mean $z + \mu t$ and variance $\sigma^2 t$, we have, for any $s > 0$, $t > 0$, and interval J,

$$
\begin{aligned}
P[Z(s+t) \in J \mid Z(s) = z] &= P[Z(t) \in J \mid Z(0) = z] \\
&= \int_J p_t(y - z)\,dy, \quad (6.4)
\end{aligned}
$$

where

$$
p_t(y) = \frac{1}{\sqrt{2\pi\sigma^2 t}} \exp\left[-\frac{(y - \mu t)^2}{2\sigma^2 t}\right] \quad (6.5)
$$

is the pdf of the normal distribution of mean μt and variance $\sigma^2 t$. Because (6.4) is the probability of transition from point z to interval J in time t, $p_t(y - z)$ or $p_t(y)$ is called the transition density of the BM $Z(t)$. Symbolically, we may write

$$
p_t(y - z) = \text{pdf of } [Z(t) = y \mid Z(0) = z]. \quad (6.6)
$$

For an SMB, the transition density is $p_t(y) = (1/\sqrt{2\pi t}) \exp[-y^2/(2t)]$.

Distribution of Brownian motion: The finite dimensional distributions of the BM $Z(t)$ are determined by the transition density. For example, for $s < t$,

$$
\begin{aligned}
&P[Z(s) \le y, Z(t) \le z \mid Z(0) = x] \\
&= \int_{-\infty}^{y} P[Z(t) \le z \mid Z(s) = u] p_s(u - x)\,du \\
&= \int_{-\infty}^{y} P[Z(t - s) \le z \mid Z(0) = u] p_s(u - x)\,du \\
&= \int_{-\infty}^{y} \int_{-\infty}^{z} p_s(u - x) p_{t-s}(v - u)\,dv\,du.
\end{aligned}
$$

In general, for $0 < t_1 < t_2 < \cdots < t_n$ and real numbers $z_0, z_1, z_2, \ldots, z_n$,

$$
\begin{aligned}
&P[Z(t_1) \le z_1, Z(t_2) \le z_2, \ldots, Z(t_n) \le z_n \mid Z(0) = z_0] \\
&= \int_{-\infty}^{z_1} \int_{-\infty}^{z_2} \cdots \int_{-\infty}^{z_n} p_{t_1}(y_1 - z_0) p_{t_2-t_1}(y_2 - y_1) \\
&\qquad \cdots p_{t_n - t_{n-1}}(y_n - y_{n-1})\,dy_n \cdots dy_2\,dy_1. \quad (6.7)
\end{aligned}
$$

Symbolically this may be written as

$$\text{joint pdf of } [Z(t_1) = z_1, Z(t_2) = z_2, \ldots, Z(t_n) = z_n \mid Z(0) = z_0]$$
$$= p_{t_1}(z_1 - z_0)p_{t_2-t_1}(z_2 - z_1)\cdots p_{t_n-t_{n-1}}(z_n - z_{n-1}). \qquad (6.8)$$

Exercise 6.1 Let $Z(t)$ be a Brownian motion with $Z(0) = z$, drift μ, and variance σ^2. Find $E[Z(s)Z(t)]$ for $s < t$.

Exercise 6.2 Let $Z(t)$ be a BM with $\mu = 2$ and $\sigma^2 = 9$, and let τ be the first time $t > 0$ when $Z(t) = 1$. Find $P[Z(\tau + 1) < 0 \mid Z(0) = 0]$.

Exercise 6.3 Let $B(t)$ be the SBM. For any real number a and $0 < t < 1$, determine the conditional distribution of $B(t)$ given $B(1) = a$. Note: The SBM $B(t)$, $0 \le t \le 1$, given $B(1) = 0$, is called a Brownian bridge process.

6.2 Standard Brownian motion and its maximum

Reflection principle: Let $B(t)$ be an SBM and let τ_a be its hitting time of point $a > 0$, which is the first time $t > 0$ when $B(t) = a$. Define

$$B^*(t) = \begin{cases} B(t), & \text{if } t \le \tau_a \\ 2a - B(t), & \text{if } t > \tau_a. \end{cases} \qquad (6.9)$$

Note that to get process $B^*(t)$, we follow the path of $B(t)$ until it reaches level a and then reflect the rest of the path about the horizontal line at this level, see Figure 6.1. It can be shown by the strong Markov property applied at the stopping time τ_a that $B^*(t)$ is also an SBM.

Maximum process of SBM: The maximum process $M(t)$ of an SBM $B(t)$ is defined by

$$M(t) = \max_{0 \le u \le t} B(u). \qquad (6.10)$$

Thus, $M(t)$ is the largest value of the SBM obtained by time t. This is depicted in Figure 6.1 for $t = 5$.

Theorem 6.2 *For $a > 0$ and $y \ge 0$,*

$$P[B(t) \le a - y, M(t) > a] = P[B(t) > a + y] = 1 - \Phi(\frac{a+y}{\sqrt{t}}), \quad (6.11)$$

where $\Phi(z) = \int_{-\infty}^{z} e^{-y^2/2} dy/\sqrt{2\pi}$ is the distribution function of the standard normal distribution.

Proof: This follows directly from the reflection principle because

$$P[B(t) \le a - y, M(t) > a] = P[B^*(t) \ge a + y]. \quad \diamondsuit$$

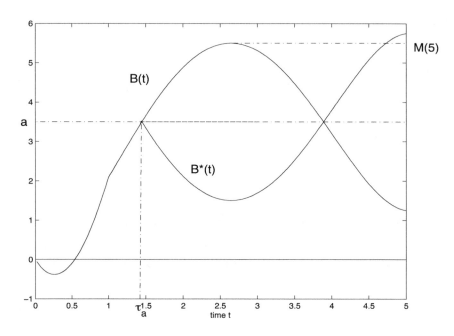

FIGURE 6.1: A reflected path.

Distribution of the maximum process $M(t)$: In Theorem 6.2, setting $y = 0$ in (6.11), we obtain the distribution of the maximum process:

$$\forall a > 0, \quad P[M(t) > a] = P[B(t) \le a, M(t) > a] + P[B(t) > a]$$
$$= 2P[B(t) > a]. \quad (6.12)$$

In particular, $M(t)$ is a continuous random variable.

Distribution of hitting time for SBM: Let τ_a be the hitting time of level $a > 0$ for an SBM. Then its distribution function is given by

$$P(\tau_a \le t) = P[M(t) \ge a] = 2P[B(t) > a] = 2[1 - \Phi(a/\sqrt{t})] \quad (6.13)$$

for $t > 0$, and hence its pdf is given by

$$f(t) = \frac{d}{dt}P(\tau_a \le t) = \frac{a}{\sqrt{2\pi t^3}}e^{-a^2/(2t)}, \quad t > 0. \tag{6.14}$$

The hitting time τ_a is a.s. finite because by (6.13),

$$P(\tau_a < \infty) = \lim_{t\to\infty} P(\tau_a \le t) = 1. \tag{6.15}$$

Joint distribution of $B(t)$ and $M(t)$: We describe here how to compute a joint probability of the form $P[B(t) \in I, M(t) \in J]$ for any two intervals I and J. First assume I and J are disjoint. If I is above J, then clearly $P[B(t) \in I, M(t) \in J] = 0$. If I is below J, say $I = (a, b]$ and $J = (c, d]$ with $a < b \le c < d$ and $c \ge 0$, then

$$P[B(t) \in I, M(t) \in J] = P[a < B(t) \le b, \, c < M(t) \le d]$$
$$= P[B(t) \le b, M(t) > c] - P[B(t) \le b, M(t) > d]$$
$$-P[B(t) \le a, M(t) > c] + P[B(t) \le a, M(t) > d]$$
$$= -\Phi(\frac{2c-b}{\sqrt{t}}) + \Phi(\frac{2d-b}{\sqrt{t}}) + \Phi(\frac{2c-a}{\sqrt{t}}) - \Phi(\frac{2d-a}{\sqrt{t}}) \tag{6.16}$$

by (6.11). For two arbitrary intervals I and J, $P[B(t) \in I, M(t) \in J]$ may be computed as a sum of probabilities of the form $P[B(t) \in I', M(t) \in J']$ for two disjoint intervals I' and J', and $P[B(t) \in J', M(t) \in J']$ for the same J', noting that for $J' = (c, d]$ with $c \ge 0$,

$$P[B(t) \in J', M(t) \in J'] = P[M(t) \in J'] - P[B(t) \le c, M(t) \in J'].$$

Example 6.3 Find $P[1 < B(4) < 3, 2 < M(4) < 3]$.

Solution: The desired probability may be written as a sum of two terms:

$$P[1 < B(4) < 2, 2 < M(4) < 3] + P[2 < B(4) < 3, 2 < M(4) < 3].$$

By (6.16), the first term is $-\Phi(1) + \Phi(2) + \Phi(1.5) - \Phi(2.5)$. The second term is

$$P[2 < M(4) < 3] - P[B(4) < 2, 2 < M(4) < 3]$$
$$= P[M(4) > 2] - P[M(4) > 3] - P[B(4) < 2, 2 < M(4) < 3]$$
$$= -2\Phi(1) + 2\Phi(1.5) - [-\Phi(1) + \Phi(2) + \Phi(\infty) - \Phi(\infty)]$$

by (6.12) and (6.16). Adding up and simplifying, and using a standard normal distribution table, we get

$$-2\Phi(1) + 3\Phi(1.5) - \Phi(2.5)$$
$$= -2(0.8413) + 3(0.9332) - 0.9938 = 0.1232.$$

Exercise 6.4 Let $Z(t)$ be a Brownian motion with $Z(0) = 0$, drift $\mu = 0$ and variance $\sigma^2 = 4$. Find:
(a) $P[Z(t)$ reaches level 2 before time 5]; and
(b) $P[Z(t)$ reaches level 2 before time 5 and is below 1 at time 5].

Exercise 6.5 Let $B(t)$ be a standard Brownian motion. Find $P[B(1) > 0.5 \mid B(t) \geq -1$ for $0 \leq t \leq 1]$.

6.3 Conditional expectation and martingales

Conditional expectation given a σ-algebra: Let X be a random variable that is either nonnegative or has a finite expectation. Recall that the conditional expectation $E(X \mid Y)$ of X given another random variable Y is a random variable Z that is a function Y, defined by

$$Z = E(X \mid Y = y) \quad \text{on } [Y = y]$$

for any possible value y of Y when Y is discrete, and in general is characterized by

$$E(Z; A) = E(X; A)$$

for any event A contained in the σ-algebra $\sigma(Y)$ generated by Y.

Recall that a σ-algebra $\mathcal{G}$ is a collection of events such that for any $A \in \mathcal{G}$, $A^c \in \mathcal{G}$ and for any sequence $A_n \in \mathcal{G}$, $\cup_n A_n \in \mathcal{G}$. A random variable Y is called $\mathcal{G}$-measurable if $[Y \leq y] \in \mathcal{G}$ for any real y.

We will define the conditional expectation $E(X \mid \mathcal{G})$ of a random variable X given $\mathcal{G}$. First assume $\mathcal{G}$ is generated by a countable partition $\{A_n\}$ of the probability space Ω, so that any event in $\mathcal{G}$ (except the empty set) is a union of events from $\{A_n\}$. Then we define

$$E(X \mid \mathcal{G}) = E(X \mid A_n) \quad \text{on } A_n$$

if $P(A_n) > 0$. This defines $E(X \mid \mathcal{G})$ up to a null set (a set of zero probability), and on this null set, we may define $E(X \mid \mathcal{G})$ to be anything. Thus, $E(X \mid \mathcal{G})$ is characterized as the unique random variable (determined up to a null set) that is $\mathcal{G}$-measurable and satisfies

$$E[E(X \mid \mathcal{G}); B] = E(X; B) \quad \text{for } B \in \mathcal{G}. \tag{6.17}$$

For a general σ-algebra $\mathcal{G}$ of events, this is taken to be the definition of $E(X \mid \mathcal{G})$. It can be shown that if $X \geq 0$ or has a finite mean, then $E(X \mid \mathcal{G})$ exists uniquely (up to a null set).

We note two extreme examples: if $\mathcal{G}$ is the σ-algebra of all events, then $E(X \mid \mathcal{G}) = X$, and if $\mathcal{G}$ is the trivial σ-algebra $\{\Omega, \emptyset\}$, then $E(X \mid \mathcal{G}) = E(X)$.

Properties of conditional expectation: In what follows, any random variable, for which a conditional expectation is taken, is assumed to have a finite mean. A random variable X is called independent of a σ-algebra $\mathcal{G}$ if $[X \leq x]$ is independent of any event in $\mathcal{G}$ for any $x \in \mathbb{R}$. The following properties are easy to verify if the σ-algebra $\mathcal{G}$ is generated by a countable partition of Ω, but hold true in general.

(a) $E(aX + bY \mid \mathcal{G}) = aE(X \mid \mathcal{G}) + bE(Y \mid \mathcal{G})$ for constants a and b.
(b) $E(XY \mid \mathcal{G}) = XE(Y \mid \mathcal{G})$ if X is $\mathcal{G}$-measurable. In particular, $E(X \mid \mathcal{G}) = X$ if X is $\mathcal{G}$-measurable.
(c) $E(X \mid \mathcal{G}) = E(X)$ if X is independent of $\mathcal{G}$.
(d) $E[E(X \mid \mathcal{G}_1) \mid \mathcal{G}_2] = E[E(X \mid \mathcal{G}_2) \mid \mathcal{G}_1] = E(X \mid \mathcal{G}_1)$ if $\mathcal{G}_1 \subset \mathcal{G}_2$. In particular,

$$E(X) = E[E(X \mid \mathcal{G})].$$

(e) $E(X \mid Y) = E[X \mid \sigma(Y)]$.

Filtration: A family of σ-algebras $\mathcal{F}_t$ of events, $t \geq 0$, is called a filtration if $\mathcal{F}_s \subset \mathcal{F}_t$ for $s < t$. The σ-algebra $\mathcal{F}_t$ may be regarded as a collection of information up to time t. Let $\mathcal{F}_\infty = \sigma(\cup_{t \geq 0} \mathcal{F}_t)$, the smallest σ-algebra containing all $\mathcal{F}_t$.

If $X(t)$ is a process, then the filtration defined by

$$\mathcal{F}_t = \sigma\{X(s); \ 0 \leq s \leq t\}$$

is called the natural filtration of the process and is denoted by $\mathcal{F}_t^X$.

A process $X(t)$ is said to be adapted to a filtration $\{\mathcal{F}_t\}$ if $X(t)$ is $\mathcal{F}_t$-measurable for any t. Any process is clearly adapted to its natural filtration. A process $X(t)$ is said to be adapted to another process $Y(t)$ if $X(t)$ is adapted to $\mathcal{F}_t^Y$.

Stopping times A random variable $\tau \geq 0$, possibly taking value ∞, is called a stopping time under a filtration $\{\mathcal{F}_t\}$, or an $\{\mathcal{F}_t\}$-stopping time, if $[\tau \leq t] \in \mathcal{F}_t$ for all $t \geq 0$. A constant time t is clearly a stopping time, and so is $\tau \wedge \sigma$ for any two stopping times τ and σ.

A stopping time of a process, as defined earlier, is just a stopping time under the natural filtration. In its extended sense as given in §1.10, the natural filtration is possibly enlarged to include some events that are independent of the process. In the sequel, a stopping time of a process means a stopping time under the natural filtration.

Martingales: Given a filtration $\{\mathcal{F}_t\}$, a right continuous process $M(t)$ is called a martingale under $\{\mathcal{F}_t\}$, or an $\{\mathcal{F}_t\}$-martingale, if it has a finite mean for any $t \geq 0$, is adapted to $\{\mathcal{F}_t\}$, and

$$E[M(t) \mid \mathcal{F}_s] = M(s) \quad \text{for } s < t. \tag{6.18}$$

One may think of $M(t)$ as the fortune of a gambler at time t, then the martingale property (6.18) means that given the information up to the present time s, the expected fortune at a future time t is equal to the present fortune.

A process may be called a martingale without specifying the filtration when the filtration is clear from the context or when the natural filtration is assumed. It is easy to see that a martingale under any filtration is also a martingale under its natural filtration.

A martingale $M(t)$ is called square integrable if $E[M(t)^2] < \infty$ for any $t \geq 0$. Using the independent increments, it is easy to show that a BM $Z(t)$ without drift is a square integrable martingale.

By property (d) of conditional expectation, if M_∞ is a random variable with a finite mean, then $M(t) = E[M_\infty \mid \mathcal{F}_t]$ is a martingale. It can be shown that it is square integrable if $E(M_\infty^2) < \infty$.

Optional stopping: The martingale property (6.18) in fact holds also for stopping times, which is called the optional stopping of martingales. We will not delve into this topic here, except to mention the following consequence of the optional stopping. If $M(t)$ is a martingale and τ is a bounded stopping time under some filtration $\mathcal{F}_t$, then

$$E[M(\tau)] = E[M(0)]. \tag{6.19}$$

This can be proved first for a bounded stopping time τ with finitely many values, and then choosing such stopping times τ_n with $\tau_n \downarrow \tau$.

6.4 Brownian motion with drift

Proposition 6.4 *(An exponential martingale)* *Let $Z(t)$ be a BM with $Z(0) = 0$, drift μ and variance σ^2. Then for any real λ,*

$$M(t) = \exp[\lambda Z(t) - (\mu\lambda + \tfrac{1}{2}\sigma^2\lambda^2)t] \qquad (6.20)$$

is a martingale with $M(0) = 1$ under the natural filtration $\mathcal{F}_t^Z$ of $Z(t)$.

Proof: By the Laplace transform of a normal distribution (see §1.8), $E(e^{\lambda Z(t)}) = \exp[(\lambda\mu + \lambda^2\sigma^2/2)t]$, and hence $E[M(t)] = 1$. By the independent and stationary increments of BM $Z(t)$, for $s < t$,

$$
\begin{aligned}
& E[M(t) \mid \mathcal{F}_s^Z] \\
= \ & M(s)E\{\exp[\lambda(Z(t) - Z(s)) + (\lambda\mu + \lambda^2\sigma^2/2)(t-s)] \mid \mathcal{F}_s^Z\} \\
= \ & M(s)E\{\exp[\lambda Z(t-s) + (\lambda\mu + \lambda^2\sigma^2/2)(t-s)]\} = M(s). \quad \diamond
\end{aligned}
$$

Theorem 6.5 *(Laplace transform of hitting time)* *Let τ_a be the hitting time of level $a > 0$ for a BM $Z(t)$ with $Z(0) = 0$, drift μ and variance σ^2. Then its Laplace transform is given by*

$$E[e^{-s\tau_a}] = \exp[-\frac{a}{\sigma^2}(\sqrt{\mu^2 + 2s\sigma^2} - \mu)] \quad \text{for } s > 0. \qquad (6.21)$$

Proof: Let $\lambda > 0$ and $s = \lambda\mu + \lambda^2\sigma^2/2$. Apply the optional stopping (6.19) to the martingale in (6.20) with $\tau = \tau_a \wedge T$ for a constant $T > 0$, $E[e^{\lambda Z(\tau_a \wedge T) - s(\tau_a \wedge T)}] = 1$. Because $e^{\lambda Z(\tau_a \wedge T) - s(\tau_a \wedge T)}$ is bounded by e^a, and as $T \uparrow \infty$, it converges to $e^{\lambda a - s\tau_a}$ if $\tau_a < \infty$ and to 0 if $\tau_a = \infty$, it follows that $E[e^{\lambda a - s\tau_a}] = 1$ noting $s > 0$. Then $E[e^{-s\tau_a}] = e^{-\lambda a}$. Solving $s = \lambda\mu + \lambda^2\sigma^2/2$ for λ yields $\lambda = (\sqrt{\mu^2 + 2s\sigma^2} - \mu)/\sigma^2$. $\quad \diamond$

Finiteness of hitting time: In (6.21), letting $s \downarrow 0$ yields

$$P(\tau_a < \infty) = \begin{cases} 1, & \text{if } \mu \geq 0 \\ \exp(-2a|\mu|/\sigma^2), & \text{if } \mu < 0. \end{cases} \qquad (6.22)$$

Thus, for $a > 0$ and $Z(0) = 0$, the hitting time τ_a is finite (with probability 1) if and only if the drift $\mu \geq 0$. When $\mu < 0$, τ_a is a deficient random variable.

Probability density of hitting time: For a BM $Z(t)$ with $Z(0) = 0$, drift μ and variance σ^2, the pdf of the hitting time τ_a of level $a > 0$ is given by

$$f(t) = \frac{a}{\sqrt{2\pi\sigma^2 t^3}} \exp[-\frac{(a - \mu t)^2}{2\sigma^2 t}] \quad \text{for } t > 0. \tag{6.23}$$

To prove this, note that by (6.14) and (6.21) with $\mu = 0$ and $\sigma^2 = 1$,

$$\int_0^\infty \frac{a}{\sqrt{2\pi t^3}} \exp[-a^2/(2t)]e^{-st} dt = \exp(-a\sqrt{2s}).$$

Replacing a and s above by a/σ^2 and $\sigma^2 s + \mu^2/2$, respectively, and then substituting $\sigma^2 t$ for t, shows that the Laplace transform of the function in (6.23) is equal to the expression on the right-hand side of (6.21).

Expectation of hitting time: Let $Z(t)$ be a BM with $Z(0) = 0$, drift μ, and variance σ^2. Then for $a > 0$,

$$E(\tau_a) = \frac{a}{\mu} \quad \text{if } \mu > 0 \quad \text{and} \quad E(\tau_a) = \infty \quad \text{if } \mu \le 0. \tag{6.24}$$

To prove this, note it is trivially true for $\mu < 0$ because then τ_a is deficient. If $\mu = 0$, by (6.23), the pdf $f(t)$ of τ_a is $\approx |a|/\sqrt{2\pi\sigma^2 t^3}$ for large t, and hence $E(\tau_a) = \int_0^\infty tf(t)dt = \infty$. If $\mu > 0$, differentiating (6.21) and then setting $s = 0$ yields $E(\tau_a) = a/\mu$.

Theorem 6.6 *(Hitting probability)* *Let $Z(t)$ be a BM with $Z(0) = 0$, drift μ, and variance σ^2. For $a > 0$ and $b > 0$, the probability that $Z(t)$ goes up to level a before going down to level $-b$ is given by*

$$P(\tau_a < \tau_{-b}) = \frac{\exp(2\mu b/\sigma^2) - 1}{\exp(2\mu b/\sigma^2) - \exp(-2\mu a/\sigma^2)} \tag{6.25}$$

for $\mu \ne 0$, and $P(\tau_a < \tau_{-b}) = b/(a + b)$ for $\mu = 0$.

Proof: Let $\tau = \tau_a \wedge \tau_{-b}$. First assume $\mu \ne 0$. Let $\lambda = -2\mu/\sigma^2$ in (6.20). The martingale $M(t)$ becomes $M(t) = \exp[\lambda Z(t)]$. By (6.22), either $\tau_a < \infty$ a.s. or $\tau_{-b} < \infty$ a.s., so $\tau < \infty$ a.s. By the optional stopping (6.19), $E[e^{\lambda Z(\tau \wedge T)}] = 1$ for constant $T > 0$. Because $e^{\lambda Z(\tau \wedge T)} \le e^{|\lambda|(a \vee b)}$, letting $T \uparrow \infty$ shows $e^{\lambda a}P(\tau_a < \tau_{-b}) + e^{-\lambda b}[1 - P(\tau_a < \tau_{-b})] = E[e^{\lambda Z(\tau)}] = 1$. Solving for $P(\tau_a < \tau_{-b})$ yields (6.25).

Now assume $\mu = 0$. Then $Z(t)$ is a martingale. By the optional stopping, $E[Z(\tau \wedge T)] = 0$. Because $Z(\tau \wedge T)$ is bounded by $a \vee b$, letting $T \uparrow \infty$, we obtain $aP(\tau_a < \tau_{-b}) - b[1 - P(\tau_a < \tau_{-b})] = E[Z(\tau)] = 0$. Solving for $P(\tau_a < \tau_{-b})$ yields $P(\tau_a < \tau_{-b}) = b/(a + b)$. ◊

Example 6.7 (geometric BM): Let $Z(t)$ be a BM with $Z(0) = 0$. A process of the form $X(t) = X(0)e^{Z(t)}$ is called a geometric BM. Later we will see that a geometric BM may be used to model the price of a stock. Let $X(t) = e^{Z(t)}$ be the price of a stock at time t with initial price $X(0) = 1$. Assume BM $Z(t)$ has drift $\mu = -2$ and variance $\sigma^2 = 4$.

(a) Find the probability that the stock price will double before it is reduced in half.

(b) Find the expected present value of the stock if it is sold when its price is doubled, assuming the interest rate is $r = 0.06$ compounded continuously.

Note: The present value of an amount x payable at time t is xe^{-rt}.

Solution: (a) Because $Z(t) = \ln X(t)$, the desired probability is

$$P(\tau_{\ln 2} < \tau_{\ln(1/2)}) = P(\tau_{\ln 2} < \tau_{-\ln 2})$$
$$= \frac{e^{2(-2)(\ln 2)/4} - 1}{e^{2(-2)(\ln 2)/4} - e^{-2(-2)(\ln 2)/4}} = 1/3.$$

(b) By (6.21), the expected present value is

$$E[2\exp(-0.06\tau_{\ln 2})] = 2\exp\{-\frac{\ln 2}{4}[\sqrt{(-2)^2 + 2(0.06)(4)} - (-2)]\} = 0.693.$$

By (6.22), $P(\tau_{\ln 2} < \infty) = \exp(-2|-2|\ln 2/4) = 1/2$, so there is a 50% chance that the stock price is never doubled.

Exercise 6.6 Let $Z(t)$ be a Brownian motion starting at 0 with drift $\mu = 5$ and variance $\sigma^2 = 9$. Find the probability that $Z(t)$ reaches level -2 before reaching 1.

Exercise 6.7 Consider an option which allows its buyer to buy a certain amount of stock at price 1 at any time. The buyer will exercise the option when the stock price reaches a certain level $x > 1$ to obtain a gain of the price difference $x - 1$. Let $X(t) = x_0 e^{Z(t)}$ be the stock price at time t, where x_0 is the price at time 0 and $Z(t)$ is a Brownian motion with $Z(0) = 0$, drift μ, and variance σ^2. Let r be the interest rate. Find the value of x that maximizes the expectation of the present value of this option.

6.5   Stochastic integrals

Irregularity of Brownian paths: Let $B(t)$ be an SBM. By definition, $B(t)$ possesses continuous paths, but it can be shown that almost all the paths are nowhere differentiable. Therefore, if $H(t)$ is another process, the integral $\int_0^t H(s)dB(t)$ cannot be defined pathwise in the usual way, either as $\int_0^t H(t)B'(t)dt$ or as a Riemann-Stieljes integral. However, $\int_0^t H(s)dB(s)$ may be defined by a different method, called Itô's stochastic integral, as will be shown in this section.

Space of square integrable random variables: A random variable X is called square integrable if $E(X^2) < \infty$. Let $\mathcal{L}^2$ be the set of all such random variables. It is a vector space.

In general, let $\mathcal{X}$ be a vector space. A number $\|X\|$ associated with each vector X in $\mathcal{X}$ is called the norm of X if for any constant a, and any two vectors X and Y,

  (a) $\|X\| \geq 0$, and $\|X\| = 0$ if and only if $X = 0$;
  (b) $\|aX\| = |a|\|X|$; and
  (c) $\|X + Y\| \leq \|X\| + \|Y\|$.

The space $\mathcal{X}$ is then called a normed space. It is easy to show that $\mathcal{L}^2$ is a normed space with norm

$$\|X\| = \sqrt{E(X^2)},$$

identifying any two random variables that are equal a.s.

In a normed space, the convergence $X_n \to X$ is defined by $\|X_n - X\| \to 0$. A sequence X_n is called a Cauchy sequence if $\|X_n - X_m\|$ can be arbitrarily small when n and m are sufficiently large, that is, given any $\varepsilon > 0$, there is an integer $N > 0$ such that $\|X_n - X_m\| < \varepsilon$ whenever $n, m \geq N$. It is easy to see that a convergent sequence is a Cauchy sequence, but a Cauchy sequence may not converge. A normed space is called complete if any Cauchy sequence converges. A complete normed space is also called a Banach space. It can be shown that $\mathcal{L}^2$ is a Banach space.

Square integrable processes: A left continuous process $H(t)$ is called square integrable if for any constant $T > 0$,

$$E[\int_0^T H(s)^2 ds] < \infty.$$

Fix $T > 0$ and let $\|H\|_T$ denote the square root of the above expression. We may consider square integral processes $H(t)$ defined only for $t \in [0, T]$. Let $B(t)$ be an SBM and let $\mathcal{L}^2_T(B)$ be the space of all square integral processes on $[0, T]$, adapted to $B(t)$. Then $\mathcal{L}^2_T(B)$ is a complete normed space under the norm $\|H\|_T$.

Simple processes: A square integrable process $H(t)$ is called simple if it is a step process, that is, there is a sequence $0 = t_0 < t_1 < t_2 < \cdots < t_n \uparrow \infty$ such that $H(t)$ is a constant on each subinterval $(t_{i-1}, t_i]$ for $i \geq 1$.

For any process $H(t)$ in $\mathcal{L}^2_T(B)$, let $\Delta: 0 = t_0 < t_1 < t_2 < \cdots < t_k = T$ be a partition of $[0, T]$ and let $H^\Delta(t)$ be the simple process defined by $H^\Delta(s) = H(t_{i-1})$ for $t_{i-1} < s \leq t_i$. By the left continuity of H, $H^\Delta \to H$ in $\mathcal{L}^2_T(B)$ as $|\Delta| \to 0$, where $|\Delta|$ is the mesh of the partition Δ, defined as the length of the longest subinterval in Δ.

Stochastic integral of a simple process: Let $H(t)$ be a simple process adapted to an SBM $B(t)$. To define the stochastic integral of $H(t)$ with respect to $B(t)$ over interval $[0, t]$, choose a partition Δ of $[0, t]$ with $t = t_k$, such that $H(t)$ is a constant on each of its subintervals, and define

$$\int_0^t H(s)dB(s) = \sum_{i=1}^k H(t_{i-1})[B(t_i) - B(t_{i-1})].$$

This integral is also denoted as $(H \cdot B)(t)$ or $(H \cdot B)_t$. Then

$$E[(H \cdot B)_t] = 0 \quad \text{and} \quad E[(H \cdot B)_t^2] = E[\int_0^t H(s)^2 ds]. \qquad (6.26)$$

By the second equality in (6.26), $\|(H \cdot B)_t\| = \|H\|_t$.

To prove (6.26), note that because H is adapted to B, $H(t_{i-1})$ is independent of $B(t_i) - B(t_{i-1})$ and hence $E[(H.B)_t] = 0$. By the independent increments of B,

$$E[(H.B)_t^2] = \sum_{i=1}^k E\{H(t_{i-1})^2[B(t_i) - B(t_{i-1}]^2\}$$

$$+ \sum_{i \neq j} E\{H(t_{i-1})H(t_{j-1})[B(t_i) - B(t_{i-1})][B(t_j) - B(t_{j-1})]\}$$

$$= \sum_{i=1}^k E[H(t_{i-1})^2]E\{[B(t_i) - B(t_{i-1})]^2\} + 0$$

$$= \sum_{i-1}^{k} E[H(t_{i-1})^2](t_i - t_{i-1}) = E[\int_0^t H(s)^2 ds].$$

Itô's stochastic integrals: We are now ready to define the stochastic integral of a general square integrable process $H(t)$ with respect to a SBM $B(t)$. Assume $H(t)$ is adapted to $B(t)$. Choose a sequence of partitions Δ_n of $[0, t]$ with $|\Delta_n| \to 0$ and let $H^n = H^{\Delta_n}$. Then

$$\|(H^m \cdot B)_t - (H^n \cdot B)_t\| = \|[(H^m - H^n) \cdot B]_t\| = \|H^m - H^n\|_t \to 0$$

as $n, m \to \infty$. This shows that $(H^n \cdot B)_t$ is a Cauchy sequence in $\mathcal{L}^2$. By the completeness of $\mathcal{L}^2$, as $n \to \infty$, $(H^n \cdot B)_t$ converges in $\mathcal{L}^2$ to a random variable in $\mathcal{L}^2$, denoted as $(H \cdot B)_t$ or $\int_0^t H(s)dB(s)$, which is called Itô's stochastic integral of $H(t)$ with respect to $B(t)$. Thus,

$$(H \cdot B)_t = \int_0^t H(s)dB(s)$$

$$= \mathcal{L}^2\text{-}\lim_{|\Delta| \to 0} \sum_{i=1}^{k} H(t_{i-1})[B(t_i) - B(t_{i-1})], \qquad (6.27)$$

where $\Delta: 0 = t_0 < t_1 < t_2 < \cdots < t_k = t$ is a sequence of partition with mesh $|\Delta| \to 0$.

Note on Itô's stochastic integrals: From (6.27), we see that Itô's stochastic integral is similar to a Riemann integral but with two differences. First, the limit is taken in $\mathcal{L}^2$ instead of pointwise, secondly, the integrand is evaluated at the left endpoint t_{i-1} of each subinterval instead of at an arbitrary point as for a Riemann integral. In fact, the limit in (6.27) may still exist when the left endpoint t_{i-1} is replaced by the right endpoint t_i, but it will be a different limit in general.

Brownian martingales: Let $B(t)$ be an SBM and let $H(t)$ be a square integrable process adapted to $B(t)$. For any $x \in \mathbb{R}$, it is not hard to show that the process

$$M(t) = x + \int_0^t H(s)dB(s) \qquad (6.28)$$

is a square integrable martingale under the Brownian filtration $\{\mathcal{F}_t^B\}$. Conversely, it can be shown that any square integrable $\{\mathcal{F}_t^B\}$-martingale $M(t)$ is continuous and admits an expression in (6.28) for some $x \in \mathbb{R}$ and some square integrable process $H(t)$ adapted to $B(t)$.

The reader is referred to a text on stochastic calculus, such as [7], for more details on this integral representation of Brownian martingales.

Stochastic integrals for more general processes: The stochastic integral may be defined for any left continuous process $H(t)$ that is adapted to the SBM $B(t)$ but not necessarily square integrable, then the $\mathcal{L}^2$-convergence has to be replaced by a weaker type of convergence, called the convergence in probability. This is defined as follows: A sequence of random variable X_n is said to converge in probability to a random variable X, denoted as $X_n \xrightarrow{P} X$, if for any $\varepsilon > 0$, there is an integer $N > 0$ such that

$$P(|X_n - X| > \varepsilon) < \varepsilon \quad \text{for } n \geq N.$$

It can be shown that both almost sure convergence and $\mathcal{L}^2$-convergence imply the convergence in probability, and the convergence in probability implies the convergence in distribution.

For a left continuous process $H(t)$ that is adapted to an SBM $B(t)$, let the simple process $H^\Delta(t)$ and its stochastic integral be defined as before. Then it can be shown that, as the mesh $|\Delta| \to 0$, the stochastic integral $(H^\Delta \cdot B)_t$ converges in probability to a random variable, which is just $(H \cdot B)_t$ as defined before when the process $H(t)$ is square integrable. This random variable will still be denoted as $(H \cdot B)_t = \int_0^t H(s)dB(s)$ and is called Itô's stochastic integral of $H(t)$ with respect to the SBM $B(t)$.

For $0 \leq a < b$, define

$$\int_a^b H(t)dB(t) = \int_0^b H(t)dB(t) - \int_0^a H(t)dB(t).$$

Basic properties of stochastic integrals: The following properties are easy to prove.

(a) $\int_0^t aH(s)dB(s) = a \int_0^t H(s)dB(s)$ for any constant a;

(b) $\int_0^t [H(s)+K(s)]dB(s) = \int_0^t H(s)dB(s) + \int_0^t K(s)dB(s)$ for any two left continuous processes $H(t)$ and $K(t)$ adapted to $B(t)$;

(c) $\int_a^c H(s)dB(s) = \int_a^b H(s)dB(s) + \int_b^c H(s)dB(s)$ for $a < b < c$; and

(d) the two equalities in (6.26) hold for any square integrable process H, that is, $E[(H \cdot B)_t] = 0$ and $E[(H \cdot B)_t^2] = E[\int_0^t H(s)^2 ds]$.

Note that the stochastic integral $(H \cdot B)_t$ as a process is continuous with $(H \cdot B)_0 = 0$.

6.6 Itô's formula and stochastic differential equations

Itô processes: Let $B(t)$ be an SBM, which is fixed throughout this section. For two processes $H(t)$ and $I(t)$ (that are left continuous and adapted to $B(t)$), the process

$$X(t) = X(0) + \int_0^t H(s)dB(s) + \int_0^t I(s)ds \qquad (6.29)$$

is called an Itô process, where the first integral above is an Itô stochastic integral and the second one is the usual Riemann integral pathwise. This may be written more concisely in a differential form:

$$dX = HdB + Idt.$$

In particular, the SBM $B(t)$ is an Itô process with $H = 1$ and $I = 0$.

Stochastic integral with respect to an Itô process: The stochastic integral of a process $J(t)$ with respect to an Itô process X given in (6.29) is defined by

$$\int_0^t J(s)dX(s) = \int_0^t J(s)H(s)dB(s) + \int_0^t J(s)I(s)ds. \qquad (6.30)$$

It is clear that $Y(t) = \int_0^t J(s)dX(s)$ is also an Itô process, and

$$dY = JdX = JHdB + JIdt.$$

Quadratic variation process: Let X and Y be two Itô processes with $dX = HdB + Idt$ and $dY = JdB + Kdt$. Define

$$\langle X, Y \rangle_t = \int_0^t H(s)J(s)ds. \qquad (6.31)$$

This is called the quadratic covariation process associated with X and Y. It can be shown that $\langle X, Y \rangle_t$ is the limit in probability of

$$\sum_{i=1}^n [X(t_i) - X(t_{i-1})][Y(t_i) - Y(t_{i-1})]$$

as the mesh of partition: $0 = t_0 < t_1 < t_2 < \cdots < t_n = t$ tends to 0.

When $X = Y$, $\langle X, X \rangle_t$ is called the quadratic variation process of $X(t)$. Note that $\langle B, B \rangle_t = t$ for an SBM $B(t)$.

Itô's formula: Let $f(t, x)$ be a function of two variables possessing continuous second-order partial derivatives, and let $X(t)$ be an Itô process. Then

$$
\begin{aligned}
f(t, X(t)) \;=\;& f(0, X(0)) + \int_0^t f_t(s, X(s))ds + \int_0^t f_x(s, X(s))dX(s) \\
&+ \frac{1}{2} \int_0^t f_{xx}(s, X(s))d\langle X, X \rangle(s), \tag{6.32}
\end{aligned}
$$

where the subscripts of f indicate the partial derivatives of f. This is called Itô's formula, which may also be written in the differential form

$$
df(t, X) = f_t(t, X)dt + f_x(t, X)dX + \frac{1}{2}f_{xx}(t, X)d\langle X, X \rangle. \tag{6.33}
$$

Note that if $dX = HdB + Idt$, then the last term in the Itô formula, $(1/2)\int_0^t f_{xx}(s, X(s))d\langle X, X \rangle(s)$, may be written as $(1/2)\int_0^t f_{xx}(s, X(s))H(s)^2 ds$. It follows that if $X(t)$ is an Itô process, then $Y(t) = f(t, X(t))$ is also an Itô process.

Outline of a proof: Let $0 = t_0 < t_1 < t_2 < \cdots < t_n = t$ be a partition of interval $[0, t]$ into n subintervals of equal length $h = t_i - t_{i-1}$ for $1 \le i \le n$. Then

$$
f(t, X(t)) = f(0, X(0)) + \sum_{i=1}^n [f(t_i, X(t_i)) - f(t_{i-1}, X(t_{i-1}))]
$$

$$
= f(0, X(0)) + \sum_{i=1}^n f_t(t_{i-1}, X(t_{i-1}))h
$$

$$
+ \sum_{i=1}^n f_x(t_{i-1}, X(t_{i-1}))[X(t_i) - X(t_{i-1})]
$$

$$
+ \frac{1}{2} \sum_{i=1}^n f_{xx}(t_{i-1}, X(t_{i-1}))[X(t_i) - X(t_{i-1})]^2 + r
$$

(by Taylor's formula, where r is the remainder)

$$
= f(0, X(0)) + J_1 + J_2 + J_3 + r.
$$

Then as $h \to 0$, $J_1 \to \int_0^t f_t(s, X(s))ds$ almost surely, and $J_2 \to \int_0^t f_x(s, X(s))dX(s)$ and the error term $r \to 0$ in probability, because

$\sum_{i=1}^{n}[X(t_i)-X(t_{i-1})]^2 \to \langle X,X\rangle_t$, $J_3 \to (1/2)\int_0^t f_{xx}(s,X(s))d\langle X,X\rangle_s$
in probability. $\diamondsuit$

Stochastic differential equations: Let $B(t)$ be an SBM, and let $a(t,x)$ and $b(t,x)$ be functions of two variables. We will consider an sde (stochastic differential equation) of the following form:

$$dX(t) = a(t,X(t))dB(t) + b(t,X(t))dt. \qquad (6.34)$$

A solution of this sde is an Itô process $X(t)$ satisfying

$$X(t) = X(0) + \int_0^t a(s,X(s))dB(s) + \int_0^t b(s,X(s))ds. \qquad (6.35)$$

White noise: If $B(t)$ has a derivative $B'(t)$, then the sde (6.34) may be formally written as $X' = a(t,X)B'+b(t,X)$. Although the derivative B' of the SBM $B(t)$ does not exist in the usual sense, it is nevertheless used in application and is called white noise. In actual computation, it always leads to a stochastic integral with respect to an SBM. In this form, the sde (6.34) may be regarded as the ordinary differential equation $X' = b(t,X)$ when a noise term $a(t,X)B'$ is added to the equation.

Existence and uniqueness of solution to sde: It can be shown that if $a(t,x)$ and $b(t,x)$ possess certain regularity properties (for example, possessing bounded derivatives), then given any $X(0)$ independent of the SBM $B(t)$, the sde (6.34) has a unique solution.

If both a and b are constants, then

$$X(t) = X(0) + aB(t) + bt$$

is clearly a solution which is a BM with drift b and variance a^2.

More generally, one may consider an sde of the form

$$dX = a(t,X)dY + b(t,X)dt$$

for a given Itô process $Y(t)$.

Stratonovich stochastic integrals: Let $X(t)$ and $Y(t)$ be two Itô processes. The Stratonovich stochastic integral of $X(t)$ with respect to $Y(t)$ is defined by

$$\int_0^t X(s)\circ dY(s) = \int_0^t X(s)dY(s) + \frac{1}{2}\langle X,Y\rangle_t, \qquad (6.36)$$

which is equal to the limit in probability of

$$\sum_{i=1}^{n} \frac{1}{2}[X(t_{i-1}) + X(t_i)][Y(t_i) - Y(t_{i-1})]$$

as the mesh of partition: $0 = t_0 < t_1 < t_2 < \cdots < t_n = t$ tends to 0.

It is sometimes useful to work with Stratonovich integrals because then Itô's formula takes the form:

$$df(t, X) = f_t(t, X)dt + f_x(t, X) \circ dX, \tag{6.37}$$

which conforms with the usual rule of calculus. Consequently, for any function $f(x)$ with an anti-derivative $F(x)$,

$$\int_0^t f(X_s) \circ dX_s = F(X_t) - F(X_0).$$

Solving an sde: It is usually impossible to obtain the solution of an sde explicitly, but if an sde can be written as $dX = a(X) \circ dY$ for a given Itô process Y_t, then it can be solved as a usual differential equation:

$$[1/a(X)] \circ dX = dY \quad \text{and} \quad \int [1/a(X)] \circ dX = Y.$$

Thus, if $\int [1/a(x)]dx = g(x)$, then $g(X_t) = Y_t$ and hence $X_t = g^{-1}(Y_t)$ is a solution of the sde. This can be verified directly using Itô's formula.

Example 6.8 Let $B(t)$ be an SBM. Consider the sde

$$dX = \nu X dt + \sigma X dB, \tag{6.38}$$

where ν and σ are two constants. This sde may be used to model the value of an investment or the price of a stock. When $\sigma = 0$, $X(t)$ satisfies the ordinary differential equation $dX = \nu X dt$, and its solution is $X(t) = X(0)e^{\nu t}$. This is the value of an investment at a constant interest rate ν compounded continuously. If $\nu < 0$, then it may be regarded as the negative value of a loan at an interest rate of $|\nu|$. When $\sigma \neq 0$, a risk is involved. One may think of $X(t)$ as the price of a stock. The constant ν is called the appreciation rate representing the trend of the stock price, whereas σ is called the volatility, representing the random risk involved.

Note that the sde (6.38) may be written as $dX = X \circ dZ$, where

$$Z(t) = \sigma B(t) + [\nu - (1/2)\sigma^2]t$$

is a BM with drift $\nu - (1/2)\sigma^2$ and variance σ^2. Then $(1/X) \circ dX = dZ$, $\ln X(t) = Z(t) + \ln X(0)$, and

$$X(t) = X(0)e^{Z(t)} = X(0)\exp[\sigma B(t) + (\nu - \frac{1}{2}\sigma^2)t] \qquad (6.39)$$

is the solution of sde (6.38), which is a geometric BM mentioned before.

Exercise 6.8 Evaluate $\int_0^t B(s)dB(s)$, where $B(t)$ is an SBM.

Exercise 6.9 (a) Evaluate $E[\cos B(t)]$.
(b) Evaluate $E[e^{2B(t)}\sin B(t)]$.
Hint: In (a), apply Itô's formula to $e^{t/2}\cos B(t)$, and in (b), note that $e^{2B}\sin B$ is the imaginary part of e^{2B+iB}.

Exercise 6.10 Solve the following stochastic differential equations:

$$\text{(a)} \quad dX = 3X^{1/3}dt + 3X^{2/3}dB, \qquad X(0) = 8,$$

$$\text{(b)} \quad dX = \sqrt{X}\,dB + (\frac{1}{4} + 2\sqrt{X})dt, \qquad X(0) = 0.$$

Exercise 6.11 Express the solution $X(t)$ of the following stochastic differential equation as a stochastic integral with respect to a standard Brownian motion $B(t)$.

$$dX = -Xdt + dB, \qquad X(0) = x_0,$$

where x_0 is a constant. Also find $E[X(t)]$ and $\text{Var}[X(t)]$.
Note: The process $X(t)$ is called an Ornstein-Uhlenbeck process. It may be regarded as a standard Brownian motion plus a drift that pushes the process back to origin at an intensity proportional to the displacement from the origin.

6.7   A single stock market model

Stock price: As in Example 6.8, the stock price $X(t)$ is assumed to satisfy the sde (6.38), $dX = \nu X dt + \sigma X dB$, where $B(t)$ is an SBM, and the appreciation rate ν and the volatility $\sigma > 0$ are constant. Solving this sde, we obtain $X(t) = X(0)e^{[\nu - (1/2)\sigma^2]t + \sigma B(t)}$ as in (6.39). The reader is referred to [6] for more details on the content of this section.

Wealth and portfolio: Let $W(t)$ be the wealth of an investor at time t. Let $S(t)$ be the amount invested in stock. The remaining amount, $W(t) - S(t)$, is invested in a bond (or deposited in bank) at a fixed rate $r > 0$ (compounded continuously). A division of the total wealth into stock and bond is called a portfolio, which is determined by $S(t)$ when $W(t)$ is known. We will allow $S(t)$ to be negative or larger than $W(t)$, that is, the short selling of stock or borrowing from the bank at the same rate r are allowed. We will assume the portfolio process $S(t)$ is square integrable.

The change in the wealth $W(t)$ is caused by the change in the stock price and the increment of bond value due to the continuous compounding of the interest. The percentage change of stock price in a small time dt is equal to $dX(t)/X(t) = \nu dt + \sigma dB(t)$, whereas the percentage change in the bond value is $d[W(t) - S(t)]/[W(t) - S(t)] = r dt$. Then

$$
\begin{aligned}
dW(t) &= r[W(t) - S(t)]dt + S(t)[\nu dt + \sigma dB(t)] \\
&= rW(t)dt + (\nu - r)S(t)dt + \sigma S(t)dB(t). \quad (6.40)
\end{aligned}
$$

Discounted value of wealth: The discounted value (the present value at $t = 0$) of the wealth $W(t)$ is $e^{-rt}W(t)$. Its change, by Itô's formula applied to $f(t, x) = e^{-rt}x$, and (6.40), is

$$
\begin{aligned}
d[e^{-rt}W(t)] &= -re^{-rt}W(t)dt + e^{-rt}dW(t) \\
&= e^{-rt}(\nu - r)S(t)dt + e^{-rt}\sigma S(t)dB(t). \quad (6.41)
\end{aligned}
$$

Let $\theta = (\nu - r)/\sigma$. The process

$$
B_0(t) = B(t) + \theta t \quad (6.42)
$$

is a BM starting at 0 with drift θ and variance 1. By (6.41), $d[e^{-rt}W(t)] = \sigma e^{-rt}S(t)dB_0(t)$. We now see that the discounted value of the wealth is

$$
e^{-rt}W(t) = W(0) + \int_0^t \sigma e^{-rs}S(s)dB_0(s). \quad (6.43)
$$

An exponential martingale: By Proposition 6.4,

$$
Z_0(t) = \exp[-\theta B(t) - \frac{1}{2}\theta^2 t] \quad (6.44)
$$

is a martingale. By the explicit expression for the pdf of an SBM, it can be shown that $e^{r|B(t)|}$ is a square integrable process for any $r > 0$, and hence $Z_0(t)$ is a square integrable martingale.

Change of the probability measure: We will now introduce another probability measure on the underlying sample space Ω. Let $\{\mathcal{F}_t\}$ be the natural filtration of the SBM $B(t)$. For any $A \in \mathcal{F}_t$, let

$$P_0(A) = E[Z_0(t); A]. \tag{6.45}$$

Because $Z_0(t)$ is a martingale with $Z_0(0) = 1$, it is easy to show that P_0 is a probability measure on $\mathcal{F}_t$ for any $t > 0$. Moreover, for $s < t$, P_0 as a probability measure on $\mathcal{F}_s$ agrees with P_0 regarded as a probability measure on $\mathcal{F}_t$ restricted to $\mathcal{F}_s$. It can be shown that P_0 can be extended to be a probability measure on $\mathcal{F}_\infty = \sigma(\cup_{t>0}\mathcal{F}_t)$. Although P_0 is in general different from the original probability measure P, they have the same null sets.

The expectation computed under P_0 will be denoted by E_0. Then

$$E_0(X) = E[XZ_0(t)] \tag{6.46}$$

for any $\mathcal{F}_t$-measurable random variable X. This clearly holds for $X = 1_A$, $A \in \mathcal{F}_t$, and then a standard monotone class argument in measure theory shows (6.46) for any $\mathcal{F}_t$-measurable X such that $E(XZ_0)$ exists.

Theorem 6.9 *Under P_0, $B_0(t)$ defined by (6.42) is an SBM. Moreover, the natural filtration of $B_0(t)$ is the same as the natural filtration $\{\mathcal{F}_t\}$ of $B(t)$.*

Proof: It is easy to see that the two processes $B(t)$ and $B_0(t)$ have the same natural filtration. To prove the theorem, it suffices to show that $B(t)$ is a BM with drift $-\theta$ and variance 1 under P_0. We first prove that $B(t)$ is a BM under P_0. For $s < t$, for any bounded Borel function f and $A \in \mathcal{F}_s$, writing $Z_0(s,t)$ for $Z_0(t)/Z_0(s) = e^{-\theta[B(t)-B(s)]-(1/2)\theta^2(t-s)}$, and using the independent and stationary increments of the SBM $B(t)$ under the original probability measure P,

$$\begin{aligned}
E_0[f(B(t) - B(s))1_A] &= E[f(B(t) - B(s))1_A Z_0(t)] \\
&= E[f(B(t) - B(s))Z_0(s,t)1_A Z_0(s)] \\
&= E[f(B(t) - B(s))Z(s,t)]E[1_A Z_0(s)] \\
&= E[f(B(t-s))Z_0(t-s)]E[1_A Z_0(s)] \\
&= E_0[f(B(t-s))]E_0(1_A).
\end{aligned}$$

This shows that $B(t) - B(s)$ is independent of $\mathcal{F}_s$ and has the same distribution as $B(t - s)$ under P_0. Therefore, $B(t)$ has independent and stationary increments, and so is a BM under P_0 by Theorem 6.1.

We now calculate the drift coefficient of $B(t)$ under P_0. Applying Itô's formula to $f(t, x) = xe^{-\theta x - (1/2)\theta^2 t}$ and noting that $f_t = -(1/2)\theta^2 f$, $f_x = (1 - \theta x)e^{-\theta x - (1/2)\theta^2 t}$, and $f_{xx} = (-2\theta + \theta^2 x)e^{-\theta x - (1/2)\sigma^2 t}$, we obtain

$$B(t)Z_0(t) = f(t, B(t)) = -\theta \int_0^t Z_0(s)ds + \int_0^t [1 - \theta B(s)]Z_0(s)dB(s).$$

By the form of the distribution density of an SBM, the integrand of the above stochastic integral is a square integrable process, and hence the stochastic integral is a martingale of mean 0. It follows that

$$E_0[B(t)] = E[B(t)Z_0(t)] = -\theta \int_0^t E[Z_0(s)]ds = -\theta t.$$

This shows that $B(t)$ has drift $-\theta$.

We now calculate $E_0[B(t)^2] = E[B(t)^2 Z_0(t)]$. Applying the Itô formula to $g(t, x) = x^2 e^{-\theta x - (1/2)\theta^2 t}$ and noting that $g_t = -(1/2)\theta^2 g$, $g_x = (2x - \theta x^2)e^{-\theta x - (1/2)\theta^2 t}$, and $g_{xx} = (2 - 4\theta x + \theta^2 x^2)e^{-\theta x - (1/2)\theta^2 t}$, we obtain

$$B(t)^2 Z_0(t) = g(t, B(t)) = \int_0^t [1 - 2\theta B(s)]Z_0(s)ds + \cdots,$$

where the omitted term is a stochastic integral of a square integrable process with respect to the SBM $B(t)$. Taking expectation yields

$$\begin{aligned} E_0[B(t)] &= E[B(t)^2 Z_0(t)] = \int_0^t E[Z_0(s)]ds - 2\theta \int_0^t E[B(s)Z_0(s)]ds \\ &= t - 2\theta \int_0^t (-\theta s)ds = t + \theta^2 t^2. \end{aligned}$$

It follows that $B(t)$ has the variance t under P_0 and hence the variance coefficient of $B(t)$ as a BM under P_0 is 1. $\diamondsuit$

Contingent claim: Fix a constant time $T > 0$. An $\mathcal{F}_T$-measurable random variable $Y \geq 0$ is called a contingent claim. We will assume $E_0(Y^2) = E[Y^2 Z_0(T)] < \infty$. Consider an option that specifies the random amount Y receivable for the buyer at time T. We are interested in finding a fair price for this option at the present time $t = 0$. To approach this problem, we will take two different points of view.

The seller of the option will receive a payment x at $t = 0$ as the price for the future obligation of paying Y at $t = T$. In order to meet this obligation, the amount x is invested in stock and bond, and it is hoped that the resulting wealth $W(T)$ at time T will be at least equal to Y. The smallest value of x, such that it is possible to choose a portfolio to satisfy $W(T) \geq Y$, is called the upper hedging price and is denoted by h_{up}.

The buyer of the option will pay x at $t = 0$ as the price for receiving a future payment Y at $t = T$. In order to at least break even, the debt $-x$ is invested in stock and bond, and it is hoped that the resulting wealth at $W(T)$ (which should be negative) plus Y will be at least equal to 0. The largest value of x, such that it is possible to choose a portfolio to satisfy $W(T) + Y \geq 0$, is called the lower hedging price and is denoted by h_{low}.

Theorem 6.10 $h_{\text{up}} = h_{\text{low}} = E_0[e^{-rT}Y]$. *Therefore, a fair price for the option is*

$$u_0 = E_0[e^{-rT}Y] = E[e^{-rT}Y Z_0(T)]. \qquad (6.47)$$

Note: The inequality $h_{\text{low}} \leq h_{\text{up}}$ holds in a more general stock market with several stocks, and random and time-dependent interest rate r, appreciation rate ν, and volatility σ (see [6]).

Proof: Let $u_0 = E_0[e^{-rT}Y]$. By (6.43), the discounted wealth process $e^{-rt}W(t)$ is a martingale under P_0. If $x > h_{\text{up}}$, then there is a portfolio such that the resulting wealth process $W(t)$ satisfies $W(0) = x$ and $W(T) \geq Y$. Thus,

$$\begin{aligned} x = W(0) &= e^{-rt}W(t)\,|_{t=0} = E_0[e^{-rT}W(T) \mid \mathcal{F}_0] \\ &= E_0[e^{-rT}W(T)] \geq E_0[e^{-rT}Y] = u_0. \end{aligned}$$

This shows that $x \geq u_0$ for any $x > h_{\text{up}}$ and hence $h_{\text{up}} \geq u_0$. On the other hand, if $x < h_{\text{low}}$, then there is a portfolio such that the resulting (negative) wealth process $W(t)$ satisfies $W(0) = -x$ and $W(T) \geq -Y$. Then

$$\begin{aligned} -x = e^{-rt}W(t)\,|_{t=0} &= E_0[e^{-rT}W(T) \mid \mathcal{F}_0] \\ &= E_0[e^{-rT}W(T)] \geq -E_0[e^{-rT}Y] = -u_0. \end{aligned}$$

This shows that $x \leq u_0$ for any $x < h_{\text{low}}$ and hence $h_{\text{low}} \leq u_0$. We have proved

$$h_{\text{low}} \leq u_0 \leq h_{\text{up}}.$$

Let $M(t) = E_0[e^{-rT}Y \mid \mathcal{F}_t]$. Then $M(t)$ is a square integrable martingale under P_0 with $M(0) = u_0$, so by the integral representation of Brownian martingale, see (6.28),

$$M(t) = u_0 + \int_0^t H(s)dB_0(s),$$

for some square integrable process H. Comparing the above with the discount wealth formula (6.43), we see that if the initial wealth is u_0 and if the portfolio is chosen according to $S(t) = e^{rt}H(t)/\sigma$, then the resulting discounted wealth process $e^{-rt}W(t)$ is just $M(t)$. Since $e^{-rT}W(T) = M(T) = E_0[e^{-rT}Y \mid \mathcal{F}_T] = e^{-rT}Y$, $W(T) = Y$. Therefore, with the initial wealth u_0, the chosen portfolio will lead to a wealth Y at time T. This means $u_0 \geq h_{\text{up}}$. A similar computation will yield $u_0 \leq h_{\text{low}}$. $\diamond$

Contingent claim depending only on the terminal stock price:
Now assume Y depends only on the price of stock at $t = T$, that is, $Y = f(X(T))$ for some Borel function $f(x)$, where $X(t)$ is the stock price process given by (6.39). Because $B(t) = B_0(t) - \theta t$ and $\theta = (\nu - r)/\sigma$,

$$X(t) = X(0)e^{(\nu-\sigma^2/2)t+\sigma B(t)} = X(0)e^{(r-\sigma^2/2)t+\sigma B_0(t)},$$

and

$$u_0 = E_0[e^{-rT}f(X(T))]$$
$$= e^{-rT}\int_{-\infty}^{\infty} f(x_0 e^{(r-\sigma^2/2)T+\sigma y})p_T(y)dy, \qquad (6.48)$$

where $x_0 = X(0)$ is the initial stock price and $p_t(y) = (2\pi t)^{-1/2}e^{-y^2/(2t)}$ is the pdf of an SBM at time t. The function $f(x)$ is assumed to satisfy a certain growth condition, such as $|f(x)| \leq a+b|x|$, for some constants $a > 0$ and $b > 0$, so that

$$E_0[f(X(T))^2] = \int_{-\infty}^{\infty} f(x_0 e^{(r-\sigma^2/2)T+\sigma y})^2 p_T(y)dy < \infty.$$

Let $W(t)$ be the wealth process given in the proof of Theorem 6.10, which starts at u_0 and yields Y at $t = T$. By Theorem 6.10, u_0 is the smallest value such that with this value as the initial wealth at time 0, there is a portfolio that yields wealth Y at time T. This implies that at any time $t \in [0, T]$, $W(t)$ is the smallest value such that with this value as wealth at time t, a portfolio exists on time interval $[t, T]$ that yields wealth Y at time T. It then follows that $W(t)$ is given by (6.48) when T and x_0 are replaced by $T - t$ and $X(t)$, respectively, that is,

$$W(t) = e^{-r(T-t)} \int_{-\infty}^{\infty} f(X(t)e^{(r-\sigma^2/2)(T-t)+\sigma y})p_{T-t}(y)dy. \qquad (6.49)$$

The process $W(t)$ represents the fair price of the option at time t and so will be called the option price process.

Optimal hedging portfolio: The portfolio process $S(t)$ associated with the option price process $W(t)$ is called the optimal hedging portfolio. It is given by

$$\begin{aligned} S(t) &= X(t)e^{-\sigma^2(T-t)/2} \int_{-\infty}^{\infty} f'(X(t)e^{(r-\sigma^2/2)(T-t)+\sigma y}) \\ &\quad e^{\sigma y}p_{T-t}(y)dy. \end{aligned} \qquad (6.50)$$

To show this, note that by (6.49) and Itô's formula,

$$\begin{aligned} dW(t) &= \{(d/dx)[e^{-r(T-t)} \int_{-\infty}^{\infty} f(xe^{(r-\sigma^2/2)(T-t)+\sigma y}) \\ &\quad p_{T-t}(y)dy]\,|_{x=X(t)}\}\sigma X(t)dB(t) + \cdots, \\ &= [e^{-r(T-t)} \int_{-\infty}^{\infty} f'(X(t)e^{(r-\sigma^2/2)(T-t)+\sigma y})e^{(r-\sigma^2/2)(T-t)+\sigma y} \\ &\quad p_{T-t}(y)dy]\sigma X(t)dB(t) + \cdots, \end{aligned}$$

where the dt-term is omitted. On the other hand, by (6.40), $dW(t) = \sigma S(t)dB(t) + \cdots$ with dt-term omitted. Comparing these two expressions for $dW(t)$ yields (6.50).

Note that in (6.50), $f(x)$ should satisfy a certain smoothness condition, for example, piecewise smooth with bounded first- and second-order derivatives, so that Itô's formula may be applied to the function $g(t, x) = e^{-r(T-t)} \int_{-\infty}^{\infty} f(xe^{(r-\sigma^2/2)(T-t)+\sigma y})p_{T-t}(y)dy$ as above.

Example 6.11 (European call option) Suppose the buyer of the option has the right to buy, at $t = T$, one share of the stock at the price

$q > 0$. The associated contingent claim is $Y = [X(T) - q]_+$. In this case, (6.48) becomes the following famous Black & Scholes formula.

$$u_0 = x_0 \Phi(\mu_+(T, x_0, q)) - qe^{-rT} \Phi(\mu_-(T, x_0, q)), \qquad (6.51)$$

where $\Phi(x) = \int_{-\infty}^{x} p_1(y)dy$ as before is the distribution function of the standard normal distribution, and

$$\mu_\pm(t, x, q) = \frac{1}{\sigma\sqrt{t}} [\ln(\frac{x}{q}) + (r \pm \frac{\sigma^2}{2})t]. \qquad (6.52)$$

To show this, let $a = \mu_-(T, x_0, q)$ and $b = \mu_+(T, x_0, q)$, and note that

$$x_0 e^{(r-\sigma^2/2)T + \sigma\sqrt{T}z} > q \quad \text{if and only if} \quad z > -a$$

and $a + \sigma\sqrt{T} = b$. We have

$$
\begin{aligned}
u_0 &= e^{-rT} \int_{-\infty}^{\infty} f(x_0 e^{(r-\sigma^2/2)T + \sigma y}) \frac{1}{\sqrt{2\pi T}} e^{-y^2/(2T)} dy \\
&= e^{-rT} \int_{-\infty}^{\infty} f(x_0 e^{(r-\sigma^2/2)T + \sigma\sqrt{T}z}) \frac{1}{\sqrt{2\pi}} e^{-z^2/2} dz \\
&= e^{-rT} \int_{-a}^{\infty} [x_0 e^{(r-\sigma^2/2)T + \sigma\sqrt{T}z} - q] \frac{1}{\sqrt{2\pi}} e^{-z^2/2} dz \\
&= x_0 \int_{-a}^{\infty} \frac{1}{\sqrt{2\pi}} e^{-(z-\sigma\sqrt{T})^2/2} dz - qe^{-rT} [1 - \Phi(-a)] \\
&= x_0 \int_{-b}^{\infty} \frac{1}{\sqrt{2\pi}} e^{-z^2/2} dz - qe^{-rT} \Phi(a) = x_0 \Phi(b) - qe^{-rT} \Phi(a).
\end{aligned}
$$

This proves (6.51). Then the option price process is

$$
\begin{aligned}
W(t) &= X(t)\Phi(\mu_+(T - t, X(t), q)) \\
&\quad - qe^{-r(T-t)} \Phi(\mu_-(T - t, X(t), q)). \qquad (6.53)
\end{aligned}
$$

Note that the expression for $\mu_\pm(t, x, q)$ makes sense only when $t > 0$. When $t = 0$, set $\mu_\pm(0, x, q) = \lim_{t\to 0} \mu_\pm(t, x, q)$. That is, $\mu_\pm(0, x, q) = \infty$ if $x > q$, $\mu_\pm(0, x, q) = -\infty$ if $x < q$, and $\mu_\pm = 0$ if $x = q$. Then $W(t)$ given in (6.53) satisfies $W(T) = [X(T) - q]_+$.

By (6.50), with $\alpha = \mu_-(T - t, X(t), q)$ and $\beta = \alpha + \sigma\sqrt{T - t} = \mu_+(T - t, X(t), q)$, the optimal hedging portfolio is

$$
\begin{aligned}
S(t) &= X(t)e^{-\sigma^2(T-t)/2} \int_{-\alpha}^{\infty} \frac{1}{\sqrt{2\pi}} e^{\sigma z\sqrt{T-t}} e^{-z^2/2} dz \\
&= X(t) \int_{-\alpha}^{\infty} \frac{1}{\sqrt{2\pi}} e^{-(z-\sigma\sqrt{T-t})^2/2} dz = X(t) \int_{-\beta}^{\infty} \frac{1}{\sqrt{2\pi}} e^{-z^2/2} dz \\
&= X(t)\Phi(\mu_+(T - t, X(t), q)). \qquad (6.54)
\end{aligned}
$$

Comparing (6.53) and (6.54) shows that $S(t) > W(t)$, hence, the optimal portfolio of the European call option always borrows.

Example 6.12 (European put option)   Suppose $Y = [q - X(T)]_+$. Then

$$u_0 = qe^{-rT}\Phi(-\mu_-(T, x_0, q)) - x_0\Phi(-\mu_+(T, x_0, q)). \qquad (6.55)$$

This is established by the same computation as in Example 6.11. To obtain the option price at time $t > 0$, one just needs to replace x_0 and T by $X(t)$ and $T - t$ in (6.55).

Bibliography

[1] Asmussen, S. (2003) "Applied probability and queues", second edition, Springer.

[2] Billingsley, P. (1986) "Probability and measure", second edition, Wiley.

[3] Flick, A. and Liao, M. (2010) "A queuing system with time varying rates", Statist. and Probab. Letters 80, pp 386-389.

[4] Gumbel, H. (1960) "Waiting lines with heterogeneous servers", Oper. Res. 8, pp 504-511.

[5] Kallenberg, O. (2002) "Foundations of modern probability", second ed., Springer-Verlag.

[6] Karatzas, I. (1997) "Lectures on the mathematics of finance", Amer. Math. Soc.

[7] Karatzas, I. and Shreve, S.E. (1991) "Brownian motion and stochastic calculus", second ed., Springer-Verlag.

[8] Massey, W.A. (1985) "Asymptotic analysis of the time dependent $M/M/1$ queue", Oper. Res. 10, pp 305-327.

[9] Resnick, S.I. (1992) "Adventures in stochastic processes", Birkhäuser.

[10] Ross, S. (1983) "Stochastic processes", John Wiley.

[11] Serfozo, R. (2009) "Basics of applied stochastic processes", Springer.

Index

$\bar{F}$, 8
inf, 4
$\liminf, \limsup$, 49
$\mathcal{L}^2$, 178
$\mathcal{L}_T^2(B)$, 179
σ-algebra, 2
$\sigma\{\cdots\}$, 20
$\overset{d}{=}$, 4
$\overset{d}{\to}$, 11
sup, 49
$a \vee b$, 69
$a \wedge b$, 31
$o(h)$, 159

a.s., 4
absorbing boundary, 92
absorbing state, 98
accessible, 97
adapted to a filtration, 173
almost sure, 4
aperiodic, 103
arrival rate, 57

balance equation, 133
BD (birth and death) process, 138
Binomial distribution, 14
Borel σ-algebra, 2, 5
Borel function, 7
Brownian motion (BM), 165
    standard (SBM), 167
bullpen discipline, 62

Chapman-Kolmogorov identity, 95, 124
class property, 103
classes of a Markov chain, 98
CLT (central limit theorem), 16
communicate, 98
conditional expectation, 22, 173
conditional probability, 3
convergence
    almost sure, 12
    in distribution for processes, 19
    in distribution for random variables, 11
convolution
    of distribution functions, 10
    of pmf's or pdf's, 11
counting process, 29
    simple, 29
current age, 68

distribution function, 4
distribution of a process, 18
dRi (direct Riemann integrable), 71
drift of a Brownian motion, 166

elementary renewal theorem, 49
embedded discrete time MC, 127
entrance time, 131
equal in distribution, 4
ergodic, 131
Erlang distribution, 16

197